REPORTING ON CLIMATE CHANGE: UNDERSTANDING THE SCIENCE

FOURTH EDITION

L. Jeremy Richardson, Editor,
with Bud Ward

THANKS TO TECHNICAL REVIEWERS & CONTRIBUTORS

The editors of this guidebook express their appreciation to David C. Bader, Ph.D., for his steadfast support of editorial independence and his adherence to the strictest standards of scientific credibility during the production of the first three editions of this guidebook.

In addition, the authors benefited significantly from outstanding technical reviews of individual chapters in prior or current editions by the following technical experts:

- Jeffrey S. Amthor, Ph.D., Office of Science, U.S. Department of Energy, Germantown, Maryland.
- David C. Bader, Ph.D., Lawrence Livermore National Laboratory, Livermore, California, formerly with DOE.
- Gerald A. Meehl, Ph.D., Senior Scientist, Climate and Global Dynamics Division, National Center for Atmospheric Research, Boulder, Colorado.
- Robert C. Worrest, Ph.D., Senior Research Scientist, Ciesin, Columbia University, New York, New York.
- Don Wuebbles, Ph.D., Head and Professor, Department of Atmospheric Science, University of Illinois, Urbana, Illinois.
- Jay Gulledge, Ph.D., Senior Scientist and Director of Science and Impacts Program, Pew Center on Global Climate Change, Arlington, Virginia.

Chapter 10 was edited and updated by Stephen O. Andersen, David W. Fahey, Marco Gonzalez, K. Madhava Sarma, Stephen Seidel, and Durwood Zaelke. Stephen O. Andersen is Co-Chair of the Montreal Protocol Technology and Economic Assessment Panel (TEAP); David Fahey is Senior Scientist at the National Oceanographic and Atmospheric Administration (NOAA) Earth System Research Laboratory, and Co-Lead Author of the WMO/UNEP Scientific Assessment Panel "Questions and Answers"; Marco Gonzalez is Executive Director of the UNEP Ozone Secretariat; K. Madhava Sarma was the founding Executive Director of the Ozone Secretariat and Senior Expert Member of TEAP; Stephen Seidel is the Vice President for Policy Analysis and General Counsel at the Pew Center on Global Climate Change; Durwood Zaelke is President of the Institute for Governance and Sustainable Development. Samira de Gobert (United Nations OzonAction Editor, Paris) provided valuable assistance on graphics. This chapter is dedicated to Sarma, who died after a short hospitalization during preparation of this manuscript. The perspectives and views presented in this chapter are those of the authors, and not necessarily the views of the organizations where they are employed.

PERMISSIONS NOTICE

The first three editions of this guidebook were produced with support from the United States Department of Energy (DOE), Office of Science, Cooperative Agreement No. DE-FG02ER63414. The contents of this document do not necessarily reflect that office's views or policies.

ISBN# 1-58576-156-2 ELI project code 021601

EDITOR'S NOTE

Considering the sheer number of advances in our understanding of climate science over the last decade, the release of this fourth edition of *Reporting on Climate Change: Understanding the Science* is long overdue. Although written primarily for journalists, this volume aims to assist educators, communicators, and the public at large.

The volume is particularly important at the present moment for at least three reasons. First, the changes in the reporting industry over the past decade are having a profound impact on science journalism. As the public moves away from print media toward online sources, the business model for news outlets has changed dramatically. In many cases, the first victims of budget cuts are—you guessed it—the science and environmental reporters. That has only exacerbated the fact that most journalists—even those assigned to covering science news—have no scientific or technical background. The need for a volume such as this has therefore only grown.

Second, the misinformation propagated by opponents of climate change action has grown more voluminous over the past year, following criticism of the Intergovernmental Panel on Climate Change (IPCC), a bunch of stolen e-mails, and an unusually snowy winter in much of the United States in 2010. As we shall see, the mistakes of the IPCC have been largely overblown, several independent reviews have cleared scientists of any scientific wrongdoing in the e-mail controversy, and the heavy snow (although allowing for easy jokes about Al Gore) is actually consistent with our understanding of climate change impacts. In short, none of these events has altered our fundamental understanding of the science of climate change. In reality, the science of human-induced climate change has become even more solid over the past decade. The third edition of this volume was based primarily on the IPCC's Third Assessment Report (TAR), released in 2001. In 2007, the IPCC released its Fourth Assessment Report (AR4), which established a stronger scientific consensus and a greater level of confidence in the conclusions.

Third, although the U.S. House of Representatives passed legislation in June 2009 to set up a cap-and-trade system to limit greenhouse gases, the 2010 elections have ushered in a new class of representatives who openly question the science of climate change, and many more aren't convinced that solving the problem is worth the cost. The political appetite on Capitol Hill for addressing climate change remains as low as ever. Even President Obama has acknowledged the reality that climate legislation will have to be tackled in "bite-size" pieces in the 112th Congress. It is critical that policy makers be armed with accurate information on the science of climate change—and the risks associated with failing to address the problem.

Finally, the editor would like to specifically recognize and thank the editor of the third edition, Bud Ward, whose talent in communicating the science of climate change was evident in that edition. The new edition relies heavily on his work, and many parts were so clear and accessible that they remain unchanged in this edition. Chapter 10 on the ozone hole was very ably updated by Stephen O. Andersen, David W. Fahey, Marco Gonzalez, K. Madhava Sarma, Stephen Seidel, and Durwood Zaelke, for whose insight and expertise we are very grateful. Special thanks to Jay Gulledge of the Pew Center on Global Climate Change, whose insightful comments and suggestions significantly improved this volume. The editors would like to offer sincere thanks to the sponsoring organization, the Environmental Law Institute, for its ongoing leadership in creating and updating this volume. Scott Schang at ELI led the effort to create the fourth edition. We would also like to thank the Department of Energy and many reviewers who made earlier editions of this guide possible.

We hope that the fourth edition of *Reporting on Climate Change: Understanding the Science* will find its way into the hands of reporters and editors alike, as they sort through the myriad dissonant voices in the public discussion on the science of climate change and what to do about it.

L. Jeremy Richardson, Ph.D.
Editor

July 2011

TABLE OF CONTENTS

EXECUTIVE SUMMARY

The fourth edition of *Reporting on Climate Change: Understanding the Science* is intended to serve as a resource for journalists who are covering the scientific and political developments on this important issue. Ultimately, the aim is to reach the general public—once armed with accurate information, journalists and editors can help foster an informed and honest policy debate.

This volume delves into the details of the important scientific concepts that underpin our understanding of climate change. These concepts are presented here in accessible language, so that those without training in science can gain insight into our knowledge of the climate system and how it works. The volume as a whole presents a complete picture of the science and serves as a reference for nontechnical readers; the main concepts and important take-home messages contained in it are summarized below.

Recent Controversies Have Not Changed Our Scientific Understanding

Most of the information here is based on the Intergovernmental Panel on Climate Change (IPCC) Fourth Assessment Report (AR4), released in 2007. The IPCC assessment process represents an ambitious scientific undertaking: the assessments are comprehensive, detailed, and consensus-based. The IPCC has an unparalleled peer-review process and is formally approved by about 120 governments as a guide for decisionmakers. It remains the gold standard in assessing and summarizing the current state of the science of climate change, despite some bad press in the past year over a small number of mistakes and some highly publicized e-mails. The AR4 (three volumes totaling almost 3,000 pages) has been scrutinized more intensely than perhaps any other scientific document in history—some 2,000 experts volunteered their time to write, edit, and review the material contained in these volumes. The AR4 is a consensus document, in that all authors of a given section had to agree to all text in that section, and all lead authors and governments had to agree to all text in the Summary for Policymakers. Consequently, the report presents a conservative or "lowest-common-denominator" interpretation of scientific confidence and certainty. (See Appendix A for an overview of the IPCC process.)

Two errors in the AR4 received considerable press attention. Both appeared in the second volume, "Impacts, Adaptation, and Vulnerability" (Working Group II). *But the fundamental scientific understanding of climate change is described in the first volume, "The Physical Science Basis," and no errors in this volume have yet surfaced.* Following is a very brief description of the controversies.

Himalayan Glaciers. The chapter on projected climate change impacts in Asia in the AR4 Working Group II volume contains an unsubstantiated and apparently implausible claim that the Himalayan glaciers could disappear by 2035. This statement was apparently based on a non-peer-reviewed paper, and its inclusion is counter to IPCC guidelines. However, this single incorrect sentence and subsequent correction do not take away from the robust and consistent statement appearing in the first volume:

> *Climate change is expected to exacerbate current stresses on water resources from population growth and economic and land-use change, including urbanisation. On a regional scale, mountain snow pack, glaciers and small ice caps play a crucial role in freshwater availability. Widespread mass losses from glaciers and reductions in snow cover over recent decades are projected to accelerate throughout the 21st century, reducing water availability, hydropower potential, and changing seasonality of flows in regions supplied by meltwater from major mountain ranges (e.g. Hindu-Kush, Himalaya, Andes), where more than one-sixth of the world population currently lives.*

Sea-Level Rise in the Netherlands. The IPCC was also criticized for allegedly overstating the threat of sea-level rise to the Netherlands. The second volume included the following incorrect sentence: "The Netherlands is an example of a country highly susceptible to both sea level rise and river flooding because 55 percent of its territory is below sea level." It turns out that the figure includes land that is above sea level but susceptible to river flooding. The figure was originally obtained directly from the Dutch government, and the agency responsible, the Netherlands Environmental Assessment Agency, recently clarified that 26% of the land area is below sea level, and another 29% is susceptible to river flooding. Although the original statement, obtained directly from the Dutch government, was incorrect, it does not undermine the IPCC's conclusions on future sea-level rise due to climate change.

In response to these errors, the IPCC and the U.N. together commissioned an independent review (http://reviewipcc.interacademycouncil.net/) of IPCC policies and procedures. The review, conducted by the InterAcademy Council, found that the IPCC's process "has been successful overall, but IPCC needs to fundamentally reform its management structure and strengthen its procedures to handle ever larger and increasingly complex climate assessments as well as the more intense public scrutiny coming from a world grappling with how best to respond to climate change." The IAC offers a number of recommendations for improving the IPCC's procedures to ensure impartiality and transparency.

Hacked E-mails. In November 2007 unknown persons hacked into the e-mail servers of the Climate Research Unit (CRU) at the University of East Anglia in England. The CRU is one of four organizations worldwide that independently gather worldwide thermometer data to construct records of global temperature. The hackers sorted through and selected more than 1,000 e-mails and posted them on the web. This action was not authorized by the owner of the e-mails (the university) and is being investigated as a possible crime.

The vast majority of the e-mails were routine and unsuspicious, if impolite, but a dozen or two created the *appearance* of controversy. To date, several independent investigations have concluded that the scientists involved in the e-mail exchanges acted ethically and properly. Unfortunately, the "controversies" reported by the press were largely phrases that had been taken out of context. For example, one researcher wrote about using a "trick" to "hide the decline." As used here, a "trick" is a clever or novel way of solving a problem—not an attempt to trick or deceive the public about climate data, as was commonly reported or implied in the press. A more in-depth description of the hacked e-mails was prepared by the Pew Center on Global Climate Change (http://www.pewclimate.org/docUploads/east-anglia-cru-hacked-emails-12-07-09.pdf). Five separate investigations (three in the United Kingdom and two in the United States) have cleared the scientists of any scientific misconduct, but several of them called for greater transparency in the handling of scientific data.

Warming is Unequivocal

The AR4 concluded that the warming of the climate system is unequivocal. The scientific community chose this word (with the blessing of world governments that scrutinized the summary line by line) because the increase in global average temperature is no longer in doubt. Scientists rarely use such conclusive language, preferring instead to provide probabilities and uncertainties to bound their results.

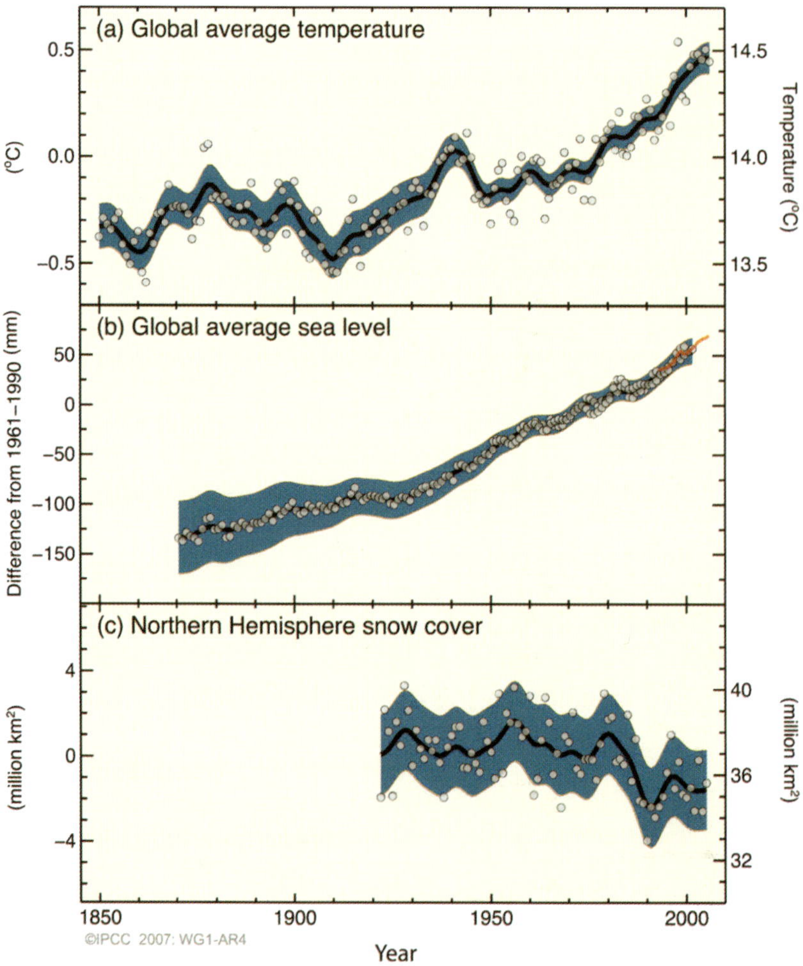

Figure 1. *Actual data showing observed changes in (a) global average surface temperature; (b) global average sea level from tide gauge measurements (blue) and satellite data (in red); and (c) Northern Hemisphere snow cover for March-April. All are differences relative to the averages for the period 1961–1990. Source: IPCC AR4, Summary for Policymakers, Figure SPM.1.*

The AR4 states in its Summary for Policymakers (WGI), *"Warming of the climate system is unequivocal, as is now evident from observations of increases in global average air and ocean temperatures, widespread melting of snow and ice, and rising global average sea level."*

Quite simply, anyone who argues that the Earth has not warmed over the past century is either misinformed or disingenuous. This question is no longer debated in the scientific community.

Observations are Consistent

The conclusion that the world has warmed is based on multiple, independent lines of evidence, which is why scientists are so convinced. Multiple data sets, observations of different physical quantities, and measurements from around the world, all point to the same conclusion—a warming world. The AR4 highlights a few of the main observational lines of evidence:

- *Both land surface air temperatures and SSTs [sea surface temperatures] show warming. In both hemispheres, land regions have warmed at a faster rate than the oceans in the past few decades, consistent with the much greater thermal inertia of the oceans.*
- *The warming of the climate is consistent with observed increases in the number of daily warm extremes, reductions in the number of daily cold extremes and reductions in the number of frost days at mid-latitudes.*
- *Changes in temperature are broadly consistent with the observed nearly worldwide shrinkage of the cryosphere.*
- *Observations of sea-level rise since 1993 are consistent with observed changes in ocean heat content and the cryosphere.*
- *Observations are consistent with physical understanding regarding the expected linkage between water vapor and temperature, and with intensification of precipitation events in a warmer world.*

That these different lines of observational evidence support one another and lead to the same conclusion strengthens scientists' confidence in the result.

Humans Are Primarily Responsible for Changes in Climate

Having established that there has been a warming trend, the next obvious question is, why? Here the AR4 is much more certain than in previous assessments:

Most of the observed increase in global average temperatures since the mid-20th century is very likely [>90% probability] due to the observed increase in anthropogenic greenhouse gas concentrations. Discernible human influences now extend to other aspects of climate, including ocean warming, continental-average temperatures, temperature extremes, and wind patterns.

Scientists have considered all potential causes of the warming trend—not only human effects like greenhouse gas emissions but also natural effects like solar variability. The warming trend, particularly in the last 50 years, is simply not consistent with changes in the Sun over that time. The science of detection and attribution is discussed in more detail in Chapter 9, and a complete discussion of natural changes in climate appears in Chapter 2.

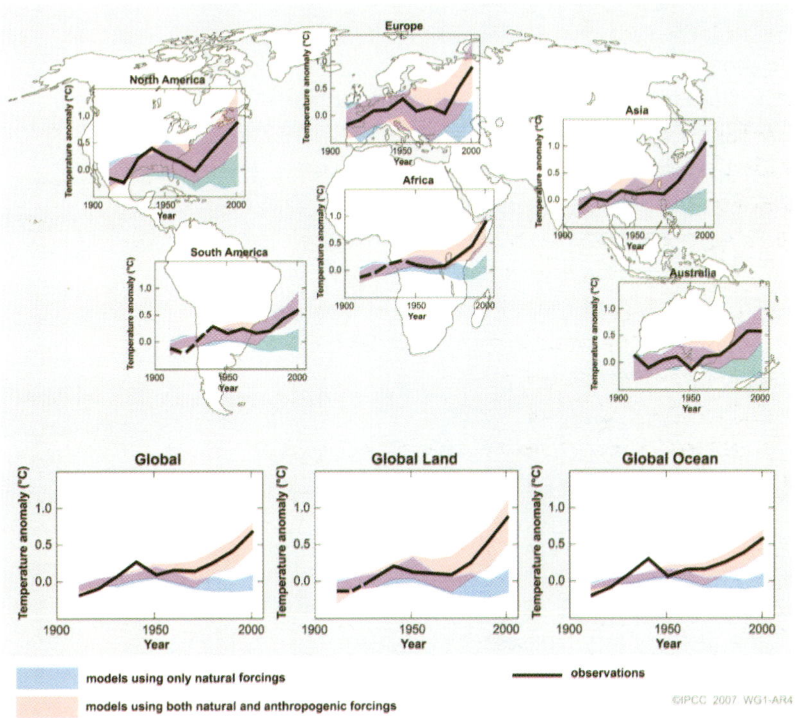

Figure 2. *Comparison of observed changes in surface temperature with climate model predictions for each continent and the globe as a whole. Data are in black, and models are shown by the shaded areas (blue for natural effects only, and red for both natural and human impacts). Source: IPCC AR4, Summary for Policymakers, Figure SPM.4.*

We Are Feeling the Effects Now

The effects of human-induced climate change are being felt around the world—including in the United States. The U.S. Global Change Research Program (USGCRP) coordinates and integrates federal research on climate change and its implications for society. Since 1989, its assessment reports have helped summarize the science of climate change (see www.globalchange.gov for helpful resources). The USGCRP released an assessment report in June 2009 called Global Climate Change Impacts in the United States, which assessed the current impacts and future projections for climate change in the United States. Among the key findings of the report are:

- ***Climate changes are underway in the United States and are projected to grow.*** *Climate-related changes are already observed in the United States and its coastal waters. These include increases in heavy downpours, rising temperature and sea level, rapidly retreating glaciers, thawing permafrost, lengthening growing seasons, lengthening ice-free seasons in the ocean and on lakes and rivers, earlier snowmelt, and alterations in river flows. These changes are projected to grow.*
- ***Widespread climate-related impacts are occurring now and are expected to increase.*** *Climate changes are already affecting water, energy, transportation, agriculture, ecosystems, and health. These impacts are different from region to region and will grow under projected climate change.*

The report can be found at http://www.globalchange.gov/publications/reports/scientific-assessments/us-impacts/key-findings.

What Kinds of Changes Are Projected?

In addition to the climate system's response to increased greenhouse gas (GHG) concentrations, many other factors will have an influence on the Earth's climate in the next century. These include variables like population growth, demographics, technological change, the extent of future GHG emissions, and policy decisions made today and beyond. For this reason, the IPCC has constructed a number of standardized socioeconomic scenarios to represent a range of plausible futures, which are contained in a separate IPCC report, the Special Report on Emission Scenarios (SRES) published in 2000. The climate models in the AR4 (and the IPCC's 2001 Third Assessment Report) were driven by the GHG emissions and land surface changes prescribed in these scenarios. Over the past decade the climate assessment research community has developed new understanding of socioeconomic drivers, and the Fifth Assessment Report, due in 2013–2014, will employ a new set of greenhouse gas emissions scenarios that are currently in development.

Even so, over the last decade our collective scientific understanding of the climate system has improved dramatically, both observationally (through longer, more numerous, and higher-quality data sets of climate variables) and theoretically (more detailed and more numerous model results suitable for comparison to observations). The AR4 was therefore able to make more highly quantitative estimates of future climate change than previous assessments. Specifically, the AR4 was able to refine ranges of expected temperature increase and sea-level rise (depending on the emissions scenario):

- As shown in the table, the global average temperature increase (above the level in 2000) projected for the year 2100 ranges from 0.6°C to 4.0°C depending on the emission scenario. Although some areas will warm more than others, an increase in temperature is expected everywhere on the globe. (See Chapter 7 for details.)
- Also shown in the table is the range of global average sea-level rise for 2100. Again, the projection is scenario-dependent, and the AR4 gives a range of 0.18 m to 0.59 m for the average increase in sea level worldwide. However, this projection is probably an underestimate, because the IPCC determined it was not possible to make an accurate prediction of future changes in the rate of ice flow from the Greenland and Antarctica ice sheets. Clearly, ice loss from the large, land-based ice sheets will be a major contributor to sea-level rise throughout this century and beyond. (See Chapter 8 for details.)
- Worldwide, snow cover is expected to decrease, along with widespread deepening of thaw depth in permafrost regions.
- Arctic sea ice is particularly sensitive to warming, and some models predict that the Arctic could be practically ice-free in the summer by the end of the century. Models for Arctic sea ice extent have underestimated the observed rate of shrinkage, and refined projections now suggest an ice-free Arctic in the summer by as early as 2035, or as late as 2080.
- Projections of changes in extreme weather events are better quantified than in previous assessments. Heat waves, for example, are expected to be more intense, longer-lasting, and more frequent in a future climate, both globally and in most regions.

Case	Temperature change (°C at 2090–2099 relative to 1980–1999)[a, d]	(m at 2090–2099 relative to 1980–1999)	Sea-level rise
	Best estimate	Likely range	Model-based range excluding future rapid dynamical changes in ice flow
Constant year 2000 concentrations[b]	0.6	0.3 – 0.9	Not available
B1 scenario	1.8	1.1 – 2.9	0.18 – 0.38
A1T scenario	2.4	1.4 – 3.8	0.20 – 0.45
B2 scenario	2.4	1.4 – 3.8	0.20 – 0.43
A1B scenario	2.8	1.7 – 4.4	0.21 – 0.48
A2 scenario	3.4	2.0 – 5.4	0.23 – 0.51
A1FI scenario	4.0	2.4 – 6.4	0.26 – 0.59

Notes:

a) Temperatures are assessed best estimates and likely uncertainty ranges from a hierarchy of models of varying complexity as well as observational constraints.

b) Year 2000 constant composition is derived from Atmosphere-Ocean General Circulation Models (AOGCMs) only.

c) All scenarios above are six SRES marker scenarios. Approximate CO_2-eq concentrations corresponding to the computed radiative forcing due to anthropogenic GHGs and aerosols in 2100 (see p. 823 of the Working Group I TAR) for the SRES B1, AIT, B2, A1B, A2, and A1FI illustrative marker scenarios are about 600, 700, 800, 850, 1250, and 1550ppm, respectively.

d) Temperature changes are expressed as the difference from the period 1980–1999. To express the change relative to the period 1850–1899 add 0.5°C.

Table 1. *Projected global average surface warming and sea-level rise at the end of the 21st century.*

Potential Consequences Are High

While it's impossible to predict with certainty the exact state of the climate a century from now, the broad projections highlighted above, and the severity of the changes, represent a few of the challenges that society will face if it fails to curb GHG emissions.

The concept of uncertainty often fuels the debate over how, or even whether, to respond to the threat of climate change. Opponents of action often argue that because we can't be sure humans are to blame, we shouldn't take economically costly actions to reduce emissions. However, policymakers are often forced to act in the face of uncertainty. Many policy decisions are probably taken in the face of less certainty than the 90% probability the IPCC has assigned to the prospect that humans are changing the climate.

Although inherent uncertainties remain (particularly with respect to human development, socioeconomics, demographics, population growth, and how sensitive the climate system is to increased GHG concentrations), it is nevertheless clear that there is significant potential for negative consequences. Evidence suggests in some cases that actual observations of climate change are outpacing model projections—meaning that we may actually be *underestimating* the potential impacts.

In reality, uncertainty cuts both ways. The average temperature of the Earth could ultimately be lower than we think, but it could also be higher. Actually, it is more likely to be higher than our best estimate at the present time. This is because the uncertainty is asymmetric—there is a *greater* chance of more serious impacts, than there is of lesser impacts. (See Figure 3.)

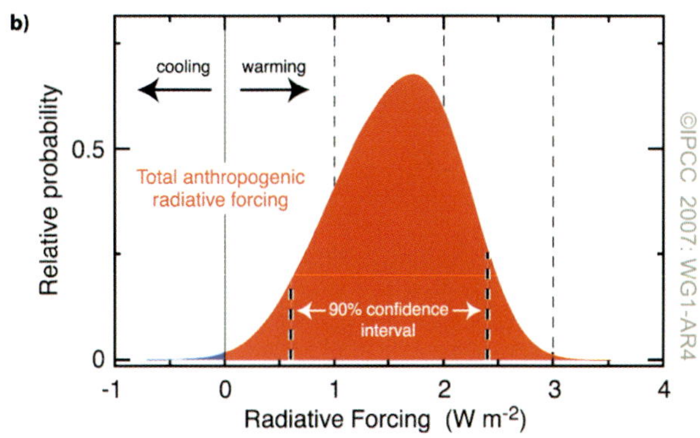

Figure 3. *This plot shows the probability of the size of the human impact on climate, in Watts per square meter (W/m²). It indicates that a range of values is possible but that the most probable value is just below 2 W/m². Importantly, negative values (meaning a cooling effect) are extremely unlikely. Source: IPCC, AR4 WGI, Technical Summary, Figure TS.5b.*

In policy circles, much discussion centers on the "safe" level of greenhouse gas concentrations in the atmosphere. 550 parts per million (ppm)? 450 ppm? 350? Or limiting the post-industrial temperature increase to 2°C? Even 1.5°C? The problem is that science can't tell us what the "safe" level is—it can only offer insights into the consequences of those physical states. In terms of making policy decisions, climate change should be considered in a risk framework, and the risks of catastrophic impacts clearly rise with the continued emissions of greenhouse gases.

Conclusion

In short, this volume provides details of the scientific principles that govern the climate system, and an in-depth look at what we do know and still don't know about climate change, all in non-technical terms. We hope this will provide editors, journalists, and the general public with critical insights into the current scientific understanding of climate change. The nature of science is such that there will always be more to learn and study, and more questions to ask. However, at the end of the day, the decision on whether to respond to the threat of climate change, and how to do so, belongs to society's decisionmakers. Ultimately, society must decide whether the benefits of responding to the threat of climate change—clean energy, new jobs, improved national security, energy savings, fewer damages due to the impacts of climate change, and reduced risks of catastrophic climatic events—are worth the costs.

CHAPTER 1:
THE CLIMATE SYSTEM AND THE FORCES THAT DRIVE IT

Climate—the prevailing regime of temperature, precipitation, humidity, wind, sunshine, snow, ice, sea conditions, etc.—is a particular characteristic of individual regions of planet Earth. The climate of the Earth as a whole is also unique—making vast areas of the Earth a hospitable refuge for life as we know it in a vast, harsh, hostile, and largely sterile universe. Carl Sagan, the famous astronomer who led NASA's Voyager missions in the 1970s and 1980s, coined the phrase "Pale Blue Dot" to describe the Earth, after the Voyager 2 spacecraft snapped a picture of the Earth from beyond Neptune.

Figure 4. *Seen from 6.1 billion kilometers (3.7 billion miles), Earth appears as a tiny dot (the blueish-white speck approximately halfway down the brown band to the right) within the darkness of deep space. Source: http://en.wikipedia.org/wiki/Pale_Blue_Dot.*

Climate on Earth has changed repeatedly as the planet has evolved geologically over its 4.6 billion-year existence. During most of the Earth's history, climate—and even the atmosphere itself—has been dramatically different from what it is today. On Earth, species have evolved and have become extinct as a result of climate change. Even during the eye blink of humans' existence on Earth, settlements, cities, and states have emerged and vanished as a result of climate change.

Understanding climate presents an enormous intellectual challenge. It involves all the "earth sciences"—physical sciences, life sciences, and some would say even social sciences. It goes way beyond meteorology (the science of weather) and beyond the atmosphere itself. Climate results from the interaction of the Sun's radiation, the Earth's orbital mechanics, the circulation and chemistry of the Earth's atmosphere, the changing polar ice and glaciers, the deep ocean currents, the weathering and shifting of the Earth's crust, and even the plants and animals that populate the Earth's surface. And now, even humans. Some geologists have begun referring to the present period in Earth's history as the "Anthropocene" to indicate that human activities are now being observed in the Earth's ecosystems.

The Sun

The energy source that drives much of what happens on Earth—not only the climate but also life itself—is the Sun. The Sun, which comprises about 99.8 percent of the mass in the solar system, consists mostly of hydrogen. At the Sun's core, hydrogen atoms are constantly being combined to form atoms of helium in the reaction known as thermonuclear fusion. In essence, the Sun is like a hydrogen bomb that has been going off continually for billions of years. Although the amount of energy released by the Sun is enormous, the intensity of solar radiation (**solar irradiance**) declines as it moves away, which is one reason why the more distant planets are colder than planets closer to the Sun, like Earth.

The Sun produces energy in the form of **electromagnetic radiation**. What is electromagnetic radiation? Technically speaking, it is a wave of fluctuating electric and magnetic fields in space. It comes in many forms, most of which people recognize. Visible light, the warmth of a glowing fire, the X-rays used by the dentist, even the radio waves we listen to—all are electromagnetic radiation.

Electromagnetic radiation travels in waves, and it is the length of these waves that distinguishes the various kinds of electromagnetic radiation. Energy coming from the Sun runs the gamut from very short-wavelength (high energy) X-rays and gamma rays, to ultraviolet and visible light, to the longer-wavelength infrared radiation (heat). But most of the energy affecting the Earth and its climate arrives at wavelengths within and near the spectrum of visible light, known to most people simply as sunlight (it is the invisible ultraviolet portion of sunlight that causes sunburn).

As solar radiation in these various wavelengths hits the Earth's atmosphere, land surface, and ocean surface, it heats them. This heat is the main energy input to our climate system. For this reason, climate scientists must—and do—study changes in the Sun's output. Of course, incoming solar energy has other effects as well. It breaks apart molecules in the atmosphere and thus drives various changes in atmospheric chemistry. Plants use solar energy to convert water and carbon dioxide into sugars and oxygen—the biological process known as **photosynthesis**. Photosynthesis is one of the foundations of the entire web of life and has shaped the Earth's atmosphere into what it is today.

For a long time, scientists believed that the amount of solar energy leaving the Sun and arriving at the Earth changed very little over time. They coined the term "solar constant" to describe something that they later realized wasn't constant—research in recent decades indicates that solar output does indeed change, over time scales ranging from minutes to decades and even longer. The solar output varies by less than 0.1 percent from decade to decade (more over longer time frames), and the value of the total, unaveraged solar irradiance hovers around 1370 Watts per square meter (W/m^2). Averaged over the surface area of the globe, the top of the Earth's atmosphere receives about 342 W/m^2 of incoming solar radiation, measured in terms of its climate-forcing effect averaged over the whole globe for an entire year. (See Forcing vs. Feedback, page 11). Small changes in this critical part of the planet's **energy budget** can have a big impact on the climate—which is why scientists are keen to understand this number and how it changes over time.

The 11-year solar cycle of sunspot activity correlates with slight changes in solar output. Other changes over longer periods have been observed or speculated upon, but high-quality, consistent modern instrumental observations have been made only since the dawn of the satellite era in the late 1970s. Historical records suggesting an absence of observed sunspots between 1645 and 1710 (the **Maunder Minimum**) have been linked with what was believed to be a cooler period of climate during the 17th to 19th centuries, known as the **Little Ice Age**. But pre-instrumental climate records are as dubious as pre-instrumental records of solar activity: recently, for example, the Little Ice Age has been pictured as a phenomenon limited to the northern hemisphere and most intense in the region surrounding the North Atlantic Ocean. (See Chapter 2 for more discussion on climate variability.)

Of the 342 W/m^2 of incoming solar radiation, some 29% is reflected back to space by clouds, the atmosphere itself, and the Earth's surface. The atmosphere

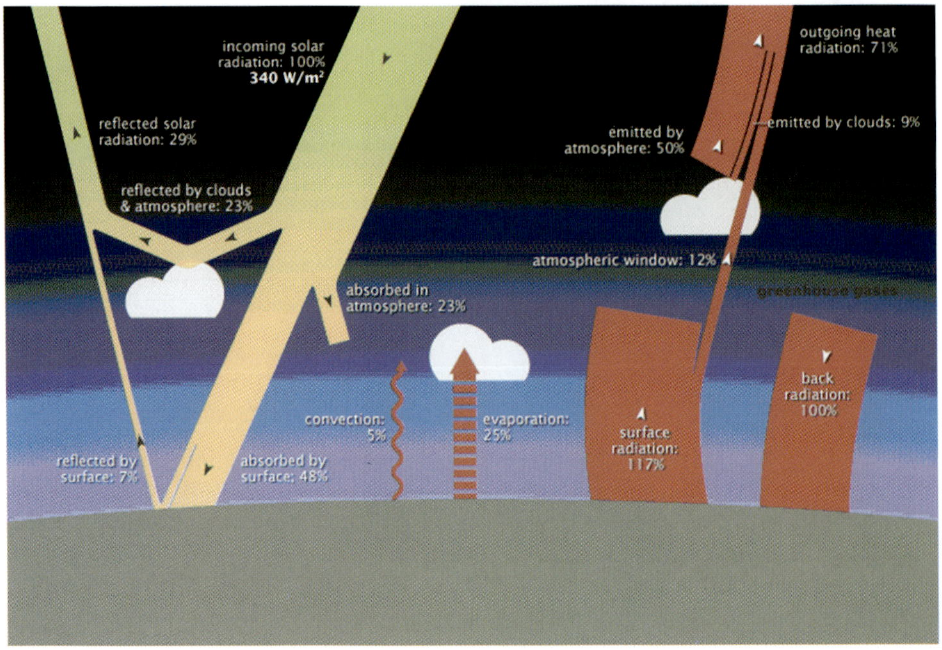

Figure 5. *The Energy Balance of the Earth's Atmosphere. The atmosphere releases heat radiation equivalent to about 59% of the incoming solar radiation, as is known from satellite measurements. This energy must be balanced by radiation absorbed by the atmosphere—but where does it come from? A large fraction comes from the incoming radiation from the Sun (about 25% of the incoming solar radiation is absorbed in the atmosphere), another large amount comes from evaporation from the Earth's surface, and the rest comes from convection and heat from the Earth's surface. Source: http://earthobservatory.nasa.gov/Features/EnergyBalance/page6.php.*

itself absorbs about 23% of the incoming radiation, and nearly half (about 48%) is absorbed by the land and ocean surface.

When incoming solar radiation, mostly in the wavelengths of visible light, is absorbed by the Earth's surface, it is converted to heat. This heat eventually warms the atmosphere in a number of ways. Some heat energy is radiated into the atmosphere as infrared radiation, the longer wavelengths we can feel when we hold our hands up to a warm bed of coals. Some of it heats the atmosphere directly via molecular motion (**convection**). Some of the heat evaporates water from the surface, causing water vapor to enter the atmosphere. The water vapor then emits heat back to the atmosphere when it rises and condenses into liquid again.

The Greenhouse Effect

Ultimately, the Earth releases back into space as much radiation energy as it receives from the Sun. Although the dominant wavelengths of incoming and outgoing radiation are different, the total amount of incoming and outgoing energy ultimately must balance as dictated by the laws of physics. In scientist-speak, the system is in **equilibrium** when this balance exists. The heated surface and atmosphere radiate their energy at infrared wavelengths. Some of it is re-radiated back down to the Earth and atmosphere, but after working its way through the system, all of this heat is eventually radiated as infrared back to outer space. The higher a body's temperature, the more heat energy it will radiate as infrared (other things being equal). If the amount of incoming energy increases, the Earth system must adjust to restore equilibrium; its surface and atmosphere will therefore warm until they reach temperatures sufficient to give off enough infrared radiation to balance the increased amount of energy the planet is taking in.

The atmosphere is transparent to most of the incoming solar radiation at the visible wavelengths in which it arrives from the Sun—it lets visible light through without absorbing much of it. However, it is less transparent to infrared and thus readily absorbs the some of the infrared radiation released from the Earth's warm surfaces.

The nitrogen and oxygen that make up about 99% of the atmosphere do not absorb infrared radiation. What absorbs it are certain **trace gases**, which each constitute only a small fraction of 1% of the entire atmosphere. The trace gases that absorb infrared radiation include water vapor, carbon dioxide, methane, nitrous oxide, ozone, and various chlorine-, fluorine-, and bromine-containing molecules.

These gases cause the atmosphere to absorb outgoing infrared energy, and to retain it longer before eventually radiating it back to space. In order to achieve the energy equilibrium that the laws of physics dictate, the Earth must warm so that the amount of infrared energy emitted back to space balances the incoming solar radiation. As a result, the temperature of the lower atmosphere overall is quite a bit warmer (about 33°C [59°F] warmer) than it would be if the atmosphere did not contain these gases. This effect is called the **greenhouse effect**—by imperfect analogy with the way the greenhouse glass generally lets in visible sunlight while trapping outgoing infrared radiation, thereby keeping the greenhouse warm. (Greenhouses also have other mechanisms in effect.) The atmosphere's greenhouse effect is entirely natural, has been well measured and documented, and is widely accepted and uncontroversial among Earth scientists.

The laws of physics govern how heat-trapping trace gases (called **greenhouse gases**, or GHGs) absorb infrared radiation. *The physical mechanism described above, by which GHGs absorb infrared radiation and heat the atmosphere, remains undisputed and has been understood for well over a century.*

In fact, the natural greenhouse effect is a big reason why the Earth is hospitable. Venus, which has large concentrations of GHGs, has an unbearable average surface temperature of 464°C, while Mercury, which is much closer to the Sun but has virtually no atmosphere to trap heat, is a cooler 167°C on average at the surface. If Earth had no atmosphere at all, it would be a frigid -18°C (0°F) on average at the Earth's surface—but the real Earth has an average surface temperature of about 15°C (59°F). Thus, the natural greenhouse effect is crucial for maintaining the average temperature at a relatively comfortable level.

However, various human activities such as fossil fuel combustion and deforestation have, since the start of the Industrial Era, been increasing atmospheric concentrations of key greenhouse gases on a global scale. On the basis of decades of scientific research, the vast majority of earth scientists now agree that the increased atmospheric concentrations of greenhouse gases are absorbing more infrared energy and causing a progressive warming of the Earth's lower atmosphere. Recent polls of scientists actively researching the climate system and Earth scientists in general support this conclusion—see Chapter 13. The portion of the warming caused by human activities is often called the **anthropogenic** greenhouse effect. (See Chapter 3, Greenhouse Gases, and Chapter 4, The Human Effect, for additional details on this discussion.) There has been considerably more controversy about the anthropogenic greenhouse effect than about the natural greenhouse effect, although the controversy is more often driven by political and economic motivations.

The following discussion will focus on some of the mechanisms that drive the climate system in addition to the Sun: atmospheric circulation, oceans, the cryosphere, and living things and changes in atmospheric composition.

Atmospheric Circulation

Incoming solar radiation warms the equator more than it does the poles. The reason is that the Sun's rays fall nearly perpendicular to the Earth's surface at the equator, but strike it obliquely (at an angle) at the poles. Not only does a given amount of solar radiation spread out over a larger land area at the poles, but it also passes through a thicker slice of the atmosphere, causing more light to be reflected back to space before reaching the surface. Moreover, the ice-covered polar surface reflects more light back to space than the ocean and land of the tropics, because ice and snow are more reflective than land, trees, and open water. (Anyone who has walked outside and squinted at snow-covered ground on a sunny day understands this intuitively.)

The natural tendency of Earth's atmosphere and oceans is to even out this heat by redistributing it from the warm equator to the frigid poles. A number of mechanisms, primarily convection, are responsible for this redistribution of heat. Convection is the tendency of warm air to rise because it is less dense, lighter, and more buoyant. This is the principle that makes chimneys draw heat and smoke upward and hot-air balloons stay aloft. A small-scale illustration of atmospheric convection is the offshore breeze many beachgoers observe each evening as the land cools more rapidly than the sea, and as rising (warm) offshore air masses pull cooler air in to replace them.

The idealized picture of a convective conveyor belt moving cool air toward the equator and warm air toward the poles is complicated by several factors. As Earth spins on its axis, the moving air is deflected by the **Coriolis effect**. This is the force that causes wind to rotate clockwise around low-pressure centers in the Northern Hemisphere and counter-clockwise in the Southern Hemisphere. Still other influences on the movement of air are exerted by the placement of landmasses and mountains. The end result is a more complex pattern of swirls and eddies.

Higher up in the atmosphere, air circulates around the planet in fairly consistent large-scale streams—such as the westerlies, the trade winds, and the jet streams. These circulation patterns determine climate at the regional scale in many places. Variations in the large-scale circulation patterns can and do occur, and these variations can often bring a major weather shift to a given region.

Oceans

Oceans, too, are a major heat-transfer engine in the Earth's climate system. Tropical oceans, especially, are big absorbers of solar energy. Oceans absorb more than half of the solar energy reaching the Earth's surface. Once the ocean surface is heated by the Sun, that heat is redistributed by a complex global system of ocean currents. The movement of ocean currents, however, is much slower than the movement of atmospheric circulation currents.

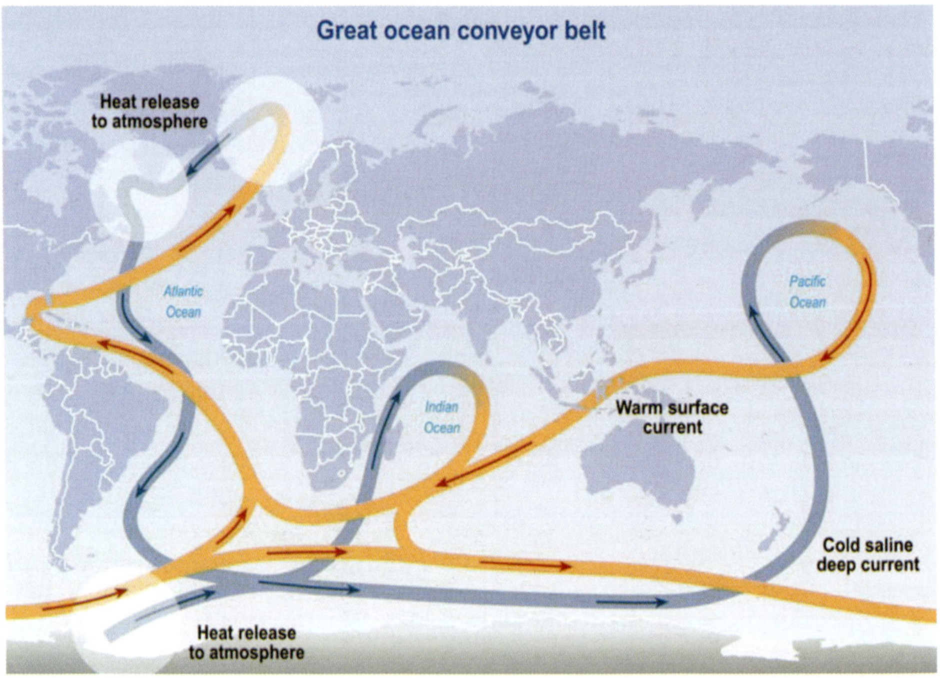

Figure 6. *Schematic illustration of the global circulation system in the world ocean consisting of major north-south thermohaline circulation routes in each ocean basin. Warm surface currents and cold deep currents are connected in the few areas of deepwater formation in the high latitudes of the Atlantic and around Antarctica (blue), where the major ocean-to-atmosphere heat transfer occurs. This current system contributes substantially to the transport and redistribution of heat. Source: IPCC, Climate Change 2001: The Scientific Basis.*

The water in the oceans is structured in layers, and these layers are important to circulation patterns. The uppermost layer, where heat is first absorbed, is well mixed by wind and waves and exchanges heat and gases with the atmosphere on the shortest timescales. Much of the energy absorbed by the ocean's upper layer evaporates water. That heat transfers to warm, humid air that produces convective updrafts that can drive weather systems. Warm late-summer surface water from the tropical Atlantic, for example, fuels hurricanes.

Currents and seasonal changes drive the mixing of water from the surface layer of the ocean with water from deeper layers. As the water mixes, there is also a transport of heat, dissolved gases, dissolved minerals, and even microscopic sea creatures. Heat convection is one of the forces that drives ocean currents, just as in the atmosphere, but there are different forces also at work. Prevailing winds, for example, can add momentum to ocean surface currents. Moreover, convection works differently in the oceans. The density differences between water masses are caused not just by temperature but also by salinity. The saltier the water is, the denser it is. Thus, cold, salty water tends to sink below warmer, fresher water. The large currents set up by this force are called **thermohaline circulation**. (Figure 6.)

Thermohaline circulation offers an example of the complexities and feedbacks involved in ocean currents, and the many ways in which they interact with surface weather and climate. When water evaporates from the surface layer, for example, the surface water becomes cooler and saltier. Precipitation and the melting of ice can make surface water fresher. In other words, activity in the atmosphere affects activity in the ocean, even as activity in the ocean affects activity in the atmosphere. The complexity of such feedbacks is one thing that makes predicting climate change difficult.

One of the best-known examples of thermohaline circulation is the Gulf Stream, a "river" of relatively warm, fresh surface water that flows from the tropics to the northern North Atlantic Ocean, where it gives off its heat to the atmosphere. As the water becomes colder and saltier, it sinks. This process is a major driver of the global thermohaline circulation of the ocean. The heat transported from the tropics northward also makes much of Western Europe considerably warmer than it otherwise would be. The notion that this North Atlantic "conveyor belt" might stop transporting heat to northern latitudes was the inspiration for the science fiction thriller, *The Day After Tomorrow.* Although the movie greatly exaggerated the potential effects of altered ocean currents, it reflected the fact that ocean heat transport is a major determinant of regional climate patterns.

The ocean surface layer both absorbs and releases carbon dioxide (CO_2), a major greenhouse gas, and oxygen, a gas essential to many forms of life. In fact, the ocean acts as a pump that removes a significant portion of the added anthropogenic CO_2 from the atmosphere. This dissolved CO_2 is taken up by microscopic plant life, phytoplankton, to make shells, and works its way through the food web as the phytoplankton are consumed by other forms of marine life. When these plankton die, their shells sift down as sediment to the ocean floor, where the CO_2 is deposited into long-term storage as calcium carbonate—fated, over geological timescales, to become limestone. The rate at which this biological carbon pump removes CO_2 from the atmosphere depends partly on how ocean currents mix surface layers and deeper layers. Of growing concern is that the increase in atmospheric concentration of CO_2 is changing the chemistry of the oceans, threatening various shell-forming organisms and the predators that feed on them.

One of the most important connections of the ocean with climate is its vast **heat capacity**. The ocean has the capacity to absorb and store far more heat than does the

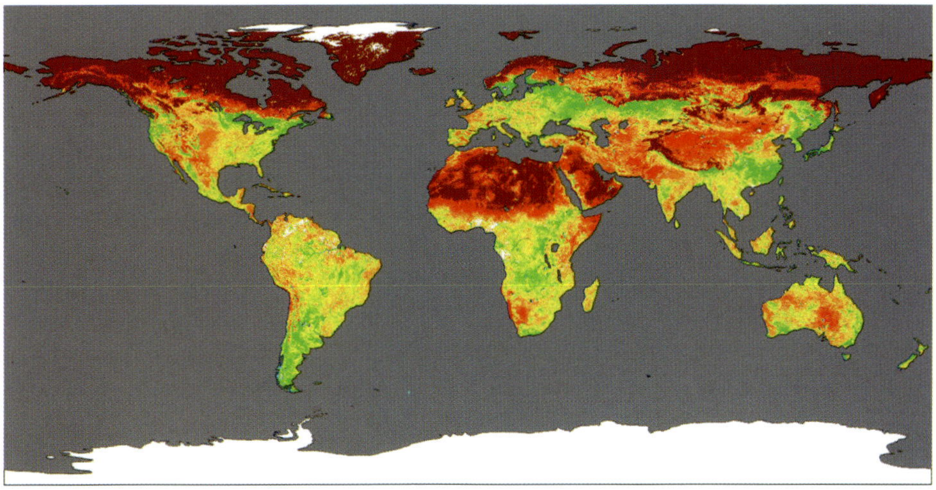

Figure 7. *How much sunlight does Earth's surface reflect? A NASA "Terra" satellite collects detailed measurements of the planet's reflectivity, its "albedo." Areas shown here in red are the brightest and most reflective regions; in yellows and greens of intermediate values; blues and violets indicate relatively dark surfaces; white indicates no data, and no data for oceans. Image drawn from data between April 7 and April 22, 2002. Source: Image courtesy Crystal Schaaf, Boston University, based upon data processed by the MODIS Land Science Team.*

atmosphere. Since it may take the oceans a long time—decades or centuries—to warm up in response to anthropogenic global warming, they are expected to cause a time lag in the response of the climate system to the forcing of an enhanced greenhouse effect.

That time lag can be viewed as both good news and bad news from a policy perspective. The good news is that it may give humans some extra time to prepare for, adapt to, or head off global warming. But the bad news is that, by delaying the appearance of observable climate change, it may lull people into a false sense of security that keeps them from recognizing the need to take necessary preventive actions. In addition, the effects of warmer surface temperatures and of rising sea levels will persist long after greenhouse gas emissions have leveled off.

Ice: The Cryosphere

Polar sea ice, continental ice sheets, seasonal and permanent sea ice, alpine glaciers, seasonal snow cover, permafrost, and other frozen aspects of the planet collectively known as the **cryosphere** also interact with climate in important ways. The most obvious and important, perhaps, is the extent to which ice and snow reflect sunlight. Scientists use the term **albedo** to refer to reflectivity—with an albedo of one being the high end of the scale (reflecting all incoming light) and zero being the low end (absorbing all light). (Figure 6.) Because of its whiteness, normal snow has an albedo close to one; dull black substances like charcoal have an albedo close to zero.

Polar ice holds a vast amount of water—at least 80% of the planet's fresh water is locked up in polar ice. The biggest share is in the Antarctic, whose ice sheets constitute more than seven million cubic miles of ice. As vast as the current polar ice masses are, they are dwarfed by the ice present on Earth during its regular glacial periods, or ice ages, which typically occur every 100,000 years or so (although the frequency of occurrence has changed over geologic time). The glacial cycles are triggered by cycles in the Earth's orbital mechanics—regular changes in its tilt, wobble, and orbit, much as you could observe in a spinning top.

However, the amplitude of glacial cycles (i.e., the amount of warming and cooling between cold ice ages and warm interglacials) is very likely amplified by the climate feedback effects produced by ice and snow. In other words, less snow and ice means a lower overall albedo for the Earth—which means that less incoming solar radiation is reflected back to space, causing the Earth to warm.

Higher temperatures in turn lead to even less ice and snow, further lowering the planet's albedo. This feedback mechanism serves to amplify the effect of human-induced warming.

This is an important, but not the only, interaction between the cryosphere and the climate system.

Much of the polar ocean surface is covered with sea ice during the winter, and the extent of sea-ice cover affects other climate variables like evaporation and heat transfer between ocean and atmosphere, which in turn can affect thermohaline circulation.

In fact, fairly good measurements using satellites, submarines, and other methods indicate that Arctic sea-ice extent and thickness have markedly decreased over the last three decades. The declines have actually outpaced model projections for ice loss. Since the release of the AR4, more recent studies have linked the decline in polar ice to human-induced global warming.

The cryosphere also adds to the potential impact of climate change on sea-level rise. As water warms, it expands—and as the oceans are heated by global warming, such **thermal expansion** alone is expected to cause significant sea-level rise (see Chapter 6). But the melting of mountain glaciers and land-based polar ice causes additional sea-level rise. However, scientists do not yet know how to predict how quickly large ice sheets will contribute to future sea-level rise. One speculative scenario involves the collapse of the West Antarctic Ice Sheet; such an event would cause sea level to rise much more and more suddenly than thermal expansion alone could. Even so, it's clear that the West Antarctic Ice Sheet, along with the Greenland Ice Sheet, have been contributing to sea-level rise over the last decade.

Two parts of the cryosphere that get much less media attention—and therefore much less public awareness—than the ice sheets are the vast expanses of permafrost in northern latitudes and the huge deposits of methane locked up in an icy, slushy form called hydrates or **clathrates**, usually out of sight on the sea floor. Global warming could cause thawing permafrost to release important amounts of methane and carbon dioxide—a feedback that could amplify the anthropogenic greenhouse effect. A slight warming of the ocean in certain places could potentially cause a rather sudden release of large amounts of gaseous methane from hydrates into the atmosphere. Either of these could cause sudden, dramatic changes in the Earth's climate.

Although scientists do not yet know how to predict whether or when potentially catastrophic ice sheet collapse or greenhouse gas release from permafrost or clathrates might happen, geologists have found evidence of similar events occurring during past warm climates. Hence, the risk of such events appears to be real.

Living Things, Atmospheric Change, and Climate Change

Not all the forces driving the climate system are purely physical. Living things play a role too. Life on Earth is a vast and complex interacting web of microbes, plants, and animals—in the sea and on land—and many of them influence the climate. In fact, marine organisms remove some 40% of anthropogenic CO_2 emissions from the atmosphere each year.

For example, plankton in the ocean pump the greenhouse gas carbon dioxide out of the atmosphere. Terrestrial plants, forests, and soils play important roles in removing carbon dioxide, storing it in carbon compounds. The world's soils are an important carbon sink; only a fraction of the carbon absorbed is later released to the atmosphere, although this may change in a warming world.

In fact, hundreds of millions of years ago, before life emerged from the ocean onto land, blue-green algae were largely responsible for turning the Earth's atmosphere from one with little oxygen and lots of carbon dioxide into one with lots of oxygen and only a trace of carbon dioxide, by "breathing in" CO_2 and "breathing out" O_2.

The biosphere (i.e., all life on Earth) also plays important roles in regulating the atmospheric concentration of another greenhouse gas—methane. (See Chapter 3, Greenhouse Gases.) Land cover—the type and extent of plants on various land areas—affects climate in a number of other ways. Vegetation directly affects albedo and also affects and is affected by soil moisture, surface water and ground water abundance, precipitation, and other aspects of climate.

Humans are also involved in many significant interactions with the climate system. While humans or their close ancestors and relatives have inhabited the planet for roughly one million years, life was harsh and human population stayed consistently much lower than today until the most recent ice age ended. It was then that the nomadic hunter-gatherer lifestyle was replaced by the development of agriculture, and climate change may have contributed to this shift.

Agriculture, in turn, caused two important developments. First, it brought about the beginning of an exponential explosion in human population. And second, it changed the vegetative cover of significant swaths of the planet's landmass—sometimes turning desert into paradise, and sometimes the reverse.

Industrialization and the scientific revolution, two uniquely human developments, also had profound effects on the planet, its atmosphere, and its climate. The concentration of humans in cities has actually created unique microclimates—urban "islands" that are hotter and rainier than surrounding areas. Industrial technology also led to explosive growth in the extraction and burning of fossil fuels: coal, oil, and natural gas. Fossil fuel combustion on a vast scale, combined with deforestation and agricultural land use, has raised atmospheric concentrations of the greenhouse gas carbon dioxide, contributing significantly to global warming.

CHAPTER 2: CLIMATE CHANGE AND NATURAL VARIABILITY

How do we know whether humans are changing the Earth's climate?

To answer that question we need to answer several other important questions. First, we need to know what the Earth's "natural" climate is—without human intervention. We need to know what climate changes are part of Earth's natural regime: what kind of change, how much, how fast, how often, and how it works. Without answering these questions, we can't know whether anything unusual is going on.

Natural Climate Change and Variability

Climate has been changing since the world began. It has changed repeatedly on most timescales we can measure, and it has changed catastrophically, far more radically than what is feared will occur in the next 200 years—although it's worth noting that modern human civilization wasn't around during those catastrophic events. Most climate change occurs on timescales far longer than a human lifetime: centuries, millennia, or millions of years.

Actually, climate changes constantly, even on timescales as short as decades or centuries.

> **THE BOTTOM LINE ON NATURAL CLIMATE VARIABILITY**
>
> - The Sun is by far the main source of energy to the Earth's climate system; it thus makes sense to compare solar changes to changes in the Earth's climate—and scientists have in fact done this extensively.
> - Although the Sun's output varies by about 0.1% over the 11-year solar cycle, the long-term trend over the past few decades appears to be flat—meaning that solar activity cannot be responsible for the warming observed in the last 50 years.
> - Since 1750, overall changes in the Sun's output are only about 0.05%, corresponding to a forcing of 0.12 W/m^2 (0.06 to 0.3 W/m^2), which is small compared to the net effect of human activities, which is 1.6 W/m^2 (0.6 to 2.4 W/m^2). Again, this means that the Sun cannot be the dominant driver of recent observed warming of the Earth's climate.

It's important to distinguish here between weather and climate. Weather is the atmospheric conditions at a given place on timescales from minutes to months. **Climate**, on the other hand, refers to the average weather conditions for a particular region over a long time span. For example, the normal (i.e., climatological mean) daily high and low temperatures that weather forecasters reference on TV news broadcasts are 30-year averages.

During most of its estimated 4.6-billion years, the Earth did not have the sort of atmosphere that could support life on land. Earth's early atmosphere probably included large amounts of carbon dioxide (CO_2), and it took billions of years for algae and other small, plantlike organisms in the seas to remove that CO_2 and replace it with enough oxygen that life could be sustained on land.

It was not until about 450 to 350 million years ago—recent history from the Earth's 4.6-billion-year perspective—that plants, insects, and finally fishlike animals came ashore. Until that point the "natural" global atmosphere and climate had been largely lethal to living things, and the ocean was a refuge of relative safety.

Consider the Carboniferous and Permian periods, a mere 345 to 270 million years ago. The Earth was a place of shallow seas and swampy lands, with a climate much warmer and wetter than today's, covered with profuse growth of giant ferns and primitive trees. It was this climate, much warmer than the most drastic projections for the foreseeable future, that allowed the Earth to store carbon from decayed vegetation as coal and oil.

The whole tenure of the human species on Earth is much shorter still—from a million to a few million years, depending on how you define human. Many anthropologists think that early climate change brought a change from forest to savanna that forced pre-human primates to come down out of the trees and "learn"—through natural selection—to stand upright.

Some Climate Change Patterns

A graph of the Earth's geological-scale climate trends looks a lot like the temperature chart of a patient with alternating bouts of fever and hypothermia. The wiggles and dips may at first seem random and disordered. Climatologists long have been looking for patterns and cycles in climate, and they have found some.

Scientists in all fields look for patterns, because patterns indicate that a given phenomenon may be understandable or even predictable. There are many kinds of climate trends and cycles, some clearly observed and understood, and others still hypothetical and unproven. For example, over the very long term, the long cooling trend that started before the Cambrian era about 570 million years ago might be partially explained by the depletion of carbon dioxide in an atmosphere once rich with it.

As we'll see later in this volume, human activity has interrupted and overwhelmed these long-term natural cycles.

Glacial Cycles

One of the most pronounced climate cycles is that of alternating ice ages and thaws—called glacial and **interglacial** periods.

We know from geological and other evidence that vast ice sheets (similar to the ones that now cover Greenland and Antarctica) once covered major parts of the North American and Eurasian continents. During the last million years—roughly the time during which humans have existed as a species—the Earth experienced about a dozen major glaciations. During the greatest of them, about 650,000 years ago, the Laurentide Ice Sheet beginning up near Hudson Bay covered what is now Chicago and points north in ice perhaps a mile thick. (Figure 7.) So much water was bound up in ice that the level of the seas was about 400 feet lower than it is currently.

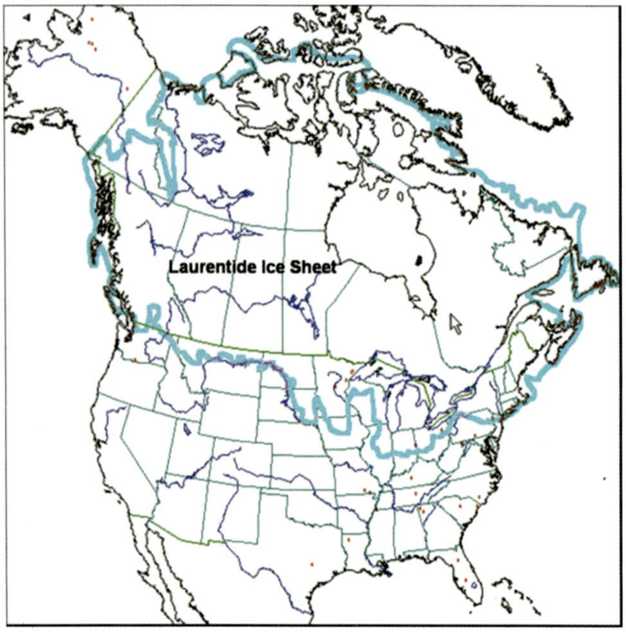

Figure 8. *Tracking the Laurentide Ice Sheet shows the retreat of glaciers in North America since the last glacial maximum 18,000 years ago. Source: NOAA Paleoclimatology Program.*

The most recent glacial period peaked about 20,000 years ago, with ice melting during the period 14,000–11,500 years ago. The glacial periods tended to last much longer than the interglacial periods. They also tended to be more pronounced in the Northern Hemisphere. While roaming bands of humans carved out a living as hunter-gatherers before that time, much of what we call human civilization, including the practice of agriculture, emerged only during the most recent interglacial period, which has (until now) been a relatively stable period in the Earth's climate.

During the most recent million years or so (the Pleistocene period), glacial periods have come at fairly regular intervals of about 100,000 years. This periodicity is pretty well explained by the **Milankovitch hypothesis**—which posits that regular wobbles and tilts in the Earth's axis of spin, and stretches in its orbit, cause changes in how warm the Earth is (especially the Northern Hemisphere where land ice comes and goes through the cycles).

The Earth's orbit stretches to become more elliptical (less round) over periods of about 100,000 years. Its axial tilt in relation to its plane of orbit changes in cycles lasting 41,000 years. The Earth's precession (the wobble of its spin, like that of a top slowing down) varies on a 23,000-year cycle.

Orbital changes alone, however, cannot fully account for the dramatic swings in the Earth's average temperature over geologic time, because the consequent changes in solar radiation reaching the Earth are very small compared to the amount of warming and cooling. Therefore, these orbital climate **forcings** must be amplified by one or more **feedback** mechanisms. A good example is the ice-albedo feedback—as temperatures warm, perhaps due to an orbital forcing, ice and snow begin to disappear globally. As the ice and snow cover decreases, much less sunlight is reflected from the Earth's surface, which amplifies the warming.

Greenhouse gases play a dramatic role in such feedback mechanisms. Analyses over the past decade of ice-core evidence have brought this role into much clearer focus. Figure 8 shows how closely temperature rises have mirrored rises in concentrations of CO_2 and methane. There are numerous physical and biological mechanisms that could cause CO_2 and methane to increase as a result of increased global temperature. It is likely, however, that interglacial warmings are caused not just by orbital mechanics, but also by greenhouse gases.

A careful examination of the data in Figure 9 shows that CO_2 increases precede temperature increases in Earth's geologic history. Contrarians have used this to argue (incorrectly) that CO_2 cannot be causing any warming. Physically, what has happened in the Earth's past is that orbital changes (Milankovitch cycles) have caused small temperature increases, leading the oceans to give off CO_2, which in turn amplifies warming. In reality, both processes occur: increased temperatures cause an increase in CO_2 concentrations, and more CO_2 in the atmosphere increases temperatures. In fact, it's clear that recent warming is being driven by GHGs like CO_2 and not orbital changes; looking at Figure 8, the difference between an ice age and an interglacial spans atmospheric CO_2 concentrations of about 180 ppm to 290 ppm. For millions of years, the concentration has stayed below about 300 ppm. Today the value is 387 ppm.

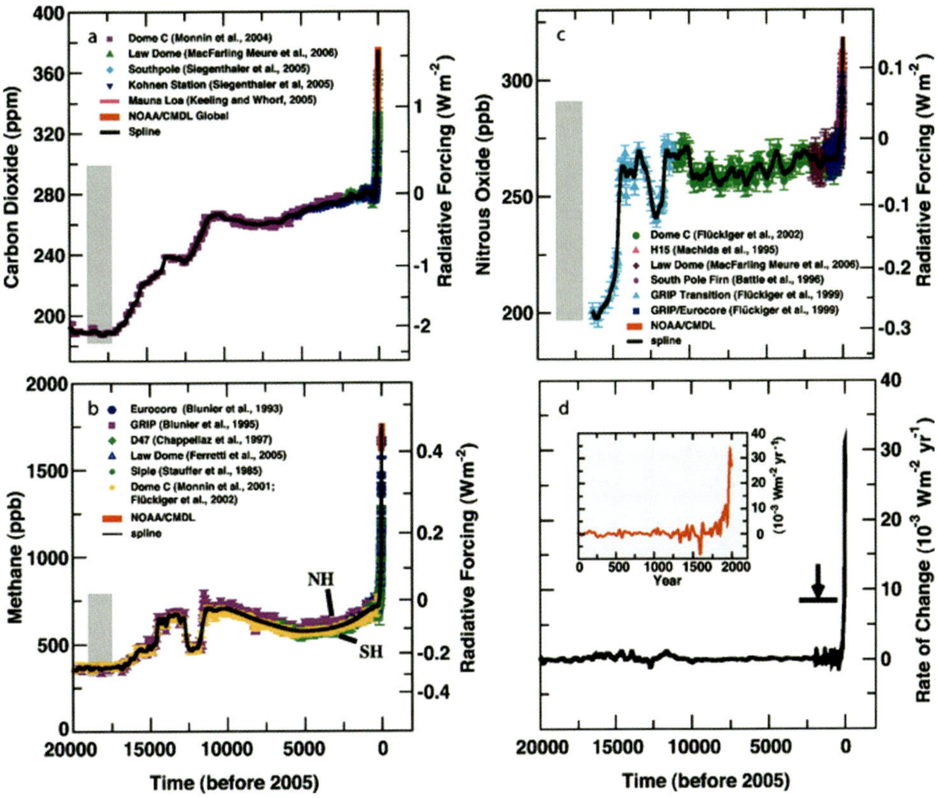

Figure 9. *These plots show the change in major GHG concentrations over the last 20,000 years. The final plot (lower right) indicates how fast the impact of this increase has changed over time (the inset zooms in on the last 2000 years). Source: IPCC, AR4 WGI, Technical Summary, Figure TS.2.*

Solar Cycles and Change

In the public discussion about climate change, few subjects are more controversial. Climate-change contrarians, many of whom operate more in the political than in

FORCING VS. FEEDBACK

Radiative forcing is a way to measure the energy balance of the Earth-atmosphere system—particularly how it changes when factors that affect climate are altered. The word **radiative** arises because these factors alter the balance between incoming radiation from the Sun and outgoing infrared radiation (heat) within the Earth's atmosphere. This balance between incoming and outgoing radiation is what controls the temperature at the Earth's surface. We use the term **forcing** to refer to a climate factor that is pushing the Earth's radiative balance away from its previous state. A positive forcing is one that causes an increase in surface temperature, while a negative forcing produces a cooling effect.

In contrast, a feedback is a factor that is not external to the climate system, but instead is affected by the prevailing climate—and can amplify the observed changes. Feedbacks do not drive changes in climate, but they have the power to amplify (potentially dramatically) the changes induced by climate forcings. Water vapor is a good example—it is a greenhouse gas with a very short atmospheric lifetime (weeks) so that it can't drive long-term changes in the Earth's radiative balance. However, as the climate gets warmer, the atmosphere can hold more water vapor, which amplifies the warming.

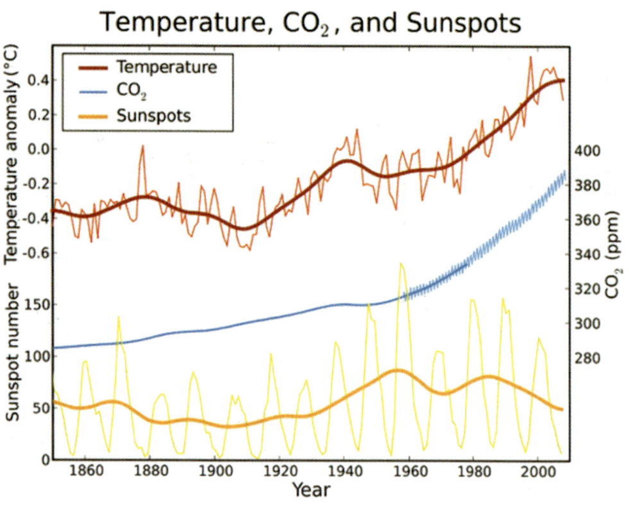

Figure 10: *Sunspots are a good proxy for solar activity. The lower curve in the plot (yellow) shows sunspot number since 1850, and the 11-year solar cycle is evident. Compared with the change in temperature and CO_2 concentrations over the same time period, it is clear that the sunspot number does not track the recent increases. Source: http://en.wikipedia.org/wiki/File%3ATemp-sunspot-co2.svg.*

the scientific arena, have seized on this issue and assert that it undermines the significance of anthropogenic greenhouse forcing (see Chapter 13). But is that really the case?

Solar output does vary, for example, with the 11-year sunspot cycle. That 11-year cyclic variation means almost nothing when it comes to the longer-term climate changes of current concern. Any warming it might cause would theoretically be reversed and undone within 11 years. What matters is whether there are any trends of increase or decrease in solar irradiance on longer timescales, trends that could cause or account for climate change.

According to the AR4, satellite measurements of solar activity over the past three decades have recorded *no significant trend*. This means that although the solar output changes by a small amount over the 11-year solar cycle, there is no evidence for a decades-long trend in solar output that could be driving changes in the Earth's climate over the last half century. New data since the TAR and improved calibration of high-quality overlapping data sets have contributed to the increased level of confidence in this conclusion.

How strong is any warming effect from solar variability likely to be in comparison with the warming effect of anthropogenic greenhouse gases? *Probably much weaker*. The AR4 put the central estimate of the warming effect of solar variability at 0.12 Watts per square meter, with a range of 0.06 to 0.3 W/m^2. This corresponds to a 0.05 percent increase in total solar irradiance since 1750. The forcing from changes in the Sun is small compared to the net effect of human activities, which is 1.6 W/m^2 (range 0.6 to 2.4 W/m^2). For a more complete discussion of sunspots, how they have changed over the past few centuries, and how these changes may have impacted the Earth's climate, see skepticalscience.com.

Ocean-Atmosphere Oscillations

While changes in an external forcing (like the Sun, described above) could have an effect on climate, changes in the climate can also come from **internal variability**. It's important to distinguish between a forcing and internal variability—the latter cannot have a long-term impact on the climate, because it is not adding energy to the system. A good example is the El Niño-Southern Oscillation (known popularly as El Niño), which periodically causes warmer temperatures in the equatorial Pacific.

Even if scientists lack a complete understanding of these effects, it is clear that they do not alter the total amount of heat in the climate system. As one area warms, another cools, resulting in no net change in global temperature. Thus, these oscillations are simply moving heat from one place to another within the climate system and cannot explain why the Earth has been warming on average for the past half century.

Other Causes of Variability

In reality, only a few mechanisms exist for altering the Earth's climate. Another example is volcanoes. When large volcanoes erupt, they can shoot huge amounts of gas, ash, and dust into the stratosphere, and some of that dust is so fine that it may take a year or more to settle out. While in the upper atmosphere, these particles shade the Earth somewhat, reflecting or absorbing solar energy that might otherwise warm the Earth's surface and lower atmosphere. This effect was demonstrated in 1816, known in New England as "The Year without a Summer," when three major eruptions took place in a short time, the largest being Tambora. More recently, the 1991 eruption of Mt. Pinatubo brought more than a year of cooler weather—and a much better understanding of the role of **aerosols** (airborne particles) in climate variability. During the Earth's geologic history, there have been periods of greater and lesser volcanism. More active periods tend to be warmer than less active periods.

So the seemingly random squiggles on Earth's temperature chart are not entirely random. Until the last few centuries, all were the result of natural phenomena, but recently, human activities have contributed. Some of the variability is forced from outside the climate system itself by things like anthropogenic greenhouse gases, solar variability, and volcanic eruptions. Other effects, like El Niño, are internal variations, which cannot be responsible for long-term changes to the climate system, since they do not add energy to the system.

As scientists have come to understand the climate system more fully in recent decades, computer models have improved, and have offered better estimates of the natural internal variability of the climate system. Of the variability that remains as yet unexplained, some may be caused by processes not yet understood. And some may be truly random or chaotic. Statisticians have rigorous tools for tackling the question of what is and is not random. When these tools are applied to climate data, it appears that some of the variation in global mean temperature from year to year is, indeed, natural, unexplained, unpredictable, and random—what scientists call **noise**.

Random variation is the kind that occurs without any pattern or order. A set of random events will each have an equal probability of occurring. Many aspects of weather really seem to be random. While gardeners in a particular location may expect the first frost of the season to occur in mid-October—whether it occurs on October 14, 15, or 16 in a given year seems purely a matter of random chance.

"Chaotic" and "random" are not exactly the same. The terms "chaos" and "chaotic" have special meanings in the context of climate and weather. During the centuries after Isaac Newton, physicists and mathematicians saw the Universe as an orderly machine that behaved according to certain well-understood laws. They believed that if they understood the operating principles of a physical system (such as the Earth's weather and climate), and knew its initial conditions, it would be possible to predict its behavior over some future period. During the 20th century, they began to understand that some mechanical systems weren't predictable—and why. You will sometimes hear such systems described as unstable or **non-linear** (meaning that "there is no simple proportional relation between cause and effect" or that small changes in initial conditions can lead to much larger changes in the final outcome).

The past decade has seen a deepened understanding of those aspects of climate that are chaotic. As computer models of the climate system have improved, they have become more useful in estimating how much chaotic climate variation could actually occur—as the same model can produce differing results from the same initial conditions. The paleoclimatic record abounds in data suggesting that large-scale warming or cooling may be more sudden, more rapid, and more dramatic than we would normally expect. The chaotic nature of climate means that surprises are possible—nasty or pleasant.

Warming Signal? Or Background Noise?

How do scientists separate the human impact from the natural changes in the climate system?

To take an imaginary example: if natural year-to-year variation in global mean temperature were only one-tenth of a degree, and the warming from people's activities were one whole degree, it would be very easy to detect. On the other hand, if the warming were a tenth of a degree and natural variation a whole degree, warming from human activities would be much more difficult to detect.

Scientists use the terms **signal** and **noise**, borrowed from information theory, to discuss this problem. If the static on a telephone line is very loud, it is hard to hear a faint voice. On the other hand, if the voice is loud and the static faint (a high signal-to-noise ratio), it is easy to understand the voice.

Two decades ago, the magnitude of the warming suspected to be a result of human activity was still not far beyond the limits of natural background variations in temperature, making a greenhouse signal difficult to detect with much certainty. That has changed. During the last two decades, the global average surface temperature has continued getting warmer and warmer. The IPCC now says that climate warming is *unequivocal*, and that 11 of the 12 years (from 1994 to 2005) rank among the top 12 warmest years on record.

The last decade has witnessed significant progress in scientific understanding of natural climate variability, both the predictable and cyclic kind and the random or chaotic kind. As the background noise and other climate signals have been more precisely described and quantified, it has become easier to distinguish the signal of human-enhanced greenhouse warming. But there is more to the story. The global warming observed during the 1990s and the first years of the 2000s was so large that it became increasingly obvious that a strengthening warming signal was emerging clearly from the noise.

CHAPTER 3:
THE BASICS OF GREENHOUSE GASES

The natural **greenhouse effect** is very real, and the basic physics is completely uncontroversial. As discussed in Chapter 1, it is a natural property of the Earth's atmosphere and a key reason the planet accommodates human life comfortably on most of its surface.

Just as light passes through the glass windows of a greenhouse, visible wavelengths of light—which make up the bulk of the solar radiation reaching the Earth—are able to pass through the atmosphere. Once the light passes through the atmosphere, it warms the land and oceans, just as it would the plants in a greenhouse. The heat from the Earth's surface then radiates outward at a longer wavelength known as **infrared** (you can feel infrared radiation if you hold your hands up to a hot stove, for example). The atmosphere, which is much less transparent to infrared radiation, traps and is heated by some of it, and eventually this extra warmth trapped in the atmosphere is radiated back to space. But in the meantime, it warms the atmosphere.

Our understanding of the basic physics leads us to predict that this will happen, and basic scientific measurements confirm it. The scientific community does not debate this phenomenon because it is so firmly rooted in the fundamental laws of physics and borne out by repeated independent observations as to be incontrovertible. Scientists know from actual measurements that the amount of solar energy absorbed by the Earth is about the same as the amount of energy radiated back into space in the form of infrared. Furthermore, scientists know that temperatures measured from satellites at the top of the atmosphere are about 60°F colder than temperatures measured at Earth's surface. This demonstrates that some of the radiation is being trapped in the atmosphere before it is reradiated back to space.

It is easy to measure how much solar or heat energy the various greenhouse gases absorb, and into what parts of the spectrum they do or don't absorb. Scientists understand the chemical composition of the atmosphere well, and absorption of infrared by the greenhouse gases is adequate to explain the heat-trapping effect.

It is important to distinguish this *natural* greenhouse effect from the potential extra warming that could be attributable to increases in the greenhouse gas content of the atmosphere as a result of human activities. It is this **anthropogenic** (human-induced) greenhouse effect that is the center of concern and debate.

Although a great deal is already known and remains solid, scientists are still researching and debating many details regarding the anthropogenic greenhouse effect, but *not* on whether the world has warmed or whether the effect is real. Scientists by and large focus their efforts on how sensitive the climate is to increased GHGs, the extent of the impacts of climate change, the timing of impacts, the regional impacts of climate change, and so on.

Climate science has matured over the past two decades. In the 1980s and 1990s, scientists argued about whether the Earth was even warming. Today we have not only modeled it, but we have been observing the

THE BOTTOM LINE ON GREENHOUSE GASES (GHGS)

- The vast majority of scientists active in climate research today are convinced that humans are the dominant driver of climate change. The AR4 clearly states, "Most of the global warming in the past 50 years is *very likely* [>90% probability] due to human increases in greenhouse gases."
- The primary greenhouse gases produced by humans are carbon dioxide (CO_2), methane (CH_4), nitrous oxide (N_2O), sulfur hexafluoride (SF_6), and two groups of compounds, hydrofluorocarbons (HFCs) and perfluorocarbons (PFCs). Collectively these GHGs are called the "Kyoto gases," after the treaty that regulates them.
- Other gases and particles can alter the energy balance of the Earth's atmosphere—these include ozone, soot, sulfates, etc.
- Water vapor is considered a feedback mechanism rather than a climate-forcing mechanism. Although it is a strong GHG, it remains in the atmosphere for only about a week or so before falling out as precipitation. As the atmosphere gets warmer, it can hold more moisture, which tends to increase warming (hence feedback).

warming—and we know without a doubt that the effect is real. Most experts who are actively engaged in climate change research today are convinced that human activity is the dominant cause of the anthropogenic greenhouse effect. The AR4 reflects this consensus in its statement that "[m]ost of the global warming in the past 50 years is *very likely* [>90% probability] due to human increases in greenhouse gases."

What's in the Air?

Almost all—99%, by volume—of the Earth's atmosphere consists of two main gases: nitrogen (about 78%) and oxygen (21%). These gases are important for a number of reasons. Oxygen, of course, is vital to the respiration of plants and animals and many microorganisms. Both gases play roles in the numerous, complex biogeochemical cycles that support life on the planet, but they play little direct role in regulating climate.

The remaining 1% or so of the Earth's atmosphere is made of small amounts of a number of **trace gases**. One of the most abundant of these is the inert gas argon, which plays no role in influencing climate. Other trace gases include water vapor, carbon dioxide, nitrous oxide, methane, chlorofluorocarbons (CFCs), and ozone—all of which can be important in the regulation of climate. These trace gases are known as the greenhouse or **radiatively active** gases (those that absorb and emit infrared radiation).

Carl Sagan once used an analogy that vividly captures the fragility of the atmosphere. He imagined the Earth to be the size of an orange and then imagined painting a coat of varnish on the orange. It turns out that the thickness of that coat of varnish compared to the size of the orange is roughly equal to the thickness of the atmosphere relative to the size of the Earth. So while the atmosphere is staggeringly big, it is actually quite thin compared to the Earth as a whole. And that thin and fragile layer is one key reason that life exists on Earth.

More specifically, because the trace gases exist in the atmosphere in such small amounts, pollution and other results of human activities can alter their proportions in quite significant ways. A change of a few parts per million in the concentration of a trace gas may represent a large percentage increase in its total atmospheric concentration (Table 2). Since trace amounts of such gases often have strong greenhouse forcing effects, the change may bring a larger increase in greenhouse forcing than we might at first imagine. Each of the gases is discussed in greater detail below.

CARBON DIOXIDE

Carbon dioxide (CO_2) is particularly important as a greenhouse gas because its greenhouse impacts are large and because human activities generate so much of it. Carbon dioxide is naturally present in the atmosphere—we only quite recently began to think of the carbon dioxide that humans add to the atmosphere as a pollutant.

What seems natural to humans today can be quite different from what is natural from the Earth's longer-term perspective, because humans have been around for only a paper-thin slice (no more than about one million years) of the Earth's 4.6-billion-year geological history. Although CO_2 makes up only about 0.039% of today's atmosphere, it was probably the dominant gas in Earth's early atmosphere. Some scientists believe that it made up as much as 80 percent of the Earth's atmosphere around 4.5 billion years ago, gradually diminishing to 30% or 20% over the next 2.5 billion years. Free oxygen was scarce to nonexistent in this early atmosphere, and indeed poisonous to most of the anaerobic life forms that existed.

Human life as we know it today would have been impossible in such a CO_2-rich atmosphere. Opening the way for humans and land animals, most of this carbon dioxide was removed from the atmosphere later in the Earth's history when sea-dwelling life, the earliest algae, evolved the ability to perform photosynthesis.

During the process of photosynthesis, plants use light energy from the Sun to turn carbon dioxide and water into sugar and oxygen. Eventually, algae—and more highly evolved organisms, like plankton, plants, and trees—died and locked up most of this carbon in the forms of carbonate minerals, oil shale, coal, and petroleum in the Earth's crust. What was left in the atmosphere is the oxygen we breathe today. Burning fossil fuels releases this ancient carbon back to the atmosphere in the form of carbon dioxide. Modern society is releasing so much carbon into the atmosphere that we are measurably changing the composition of the Earth's atmosphere.

Atmospheric carbon dioxide comes from many sources, but is usually brought into balance with **sinks** that drain carbon out of the atmosphere. Figure 10 roughly charts these carbon flows.

One of the biggest carbon **sources** is the exchange of gas between the atmosphere and the ocean surface. This exchange is actually a finely balanced, two-way process, involving tremendous amounts of carbon dioxide. Scientists measure such large amounts of carbon dioxide in gigatonnes (a billion metric tons). Often these amounts

Greenhouse Gas	Chemical Formula	Anthropogenic Sources	Atmospheric Lifetime[1] (years)	GWP[2] (100-Year Time Horizon)
Carbon Dioxide	CO_2	Fossil-fuel combustion, Land-use conversion, Cement Production	variable[1]	1
Methane	CH_4	Fossil fuels, Rice paddies, Waste dumps	12[1]	23
Nitrous Oxide	N_2O	Fertilizer, Industrial processes, Combustion	114[1]	296
CFC-12	CCL_2F_2	Liquid coolants, Foams	100	10600
HCFC-22	CCl_2F_2	Refrigerants	11.9	1700
Perfluoroethane	C_2F_6	Aluminum smelting, Semiconductor manufacturing	10000	11900
Sulfur Hexaflouride	SF_6	Dielectric fluid	3200	22200

	Pre-1750 Tropospheric Concentration[3] (parts per billion)	Current Tropospheric Concentration[4] (parts per billion)
Carbon Dioxide	280000[5,6,7]	377700[6]
Methane	730 / 688[7]	1847 / 1730[8]
Nitrous Oxide	270[7,9]	319 / 318[8]
CFC-12	0	.545 / .542[8]
HCFC-22	0	.174 / .155[8]
Perfluoroethane	0	.003[9]
Sulfur Hexaflouride	0	.00522

1 The atmospheric lifetime of carbon dioxide is difficult to define because it is exchanged with reservoirs having a wide range of turnover times; IPCC (2001) gives a range of 5–200 years. In contrast, most CH4 is removed from the atmosphere by a single process, oxidation by the hydroxyl radical (OH). The atmospheric lifetime of a gas is relatively easy to define when essentially all of its removal from the atmosphere involves a single process. However, some complications still arise. For example, the effect of an increase in atmospheric concentration of CH_4 is to reduce the OH concentration, which, in turn, reduces destruction of the additional methane, effectively lengthening its atmospheric lifetime. An opposite sort of feedback applies to N_2O: an increase induces chemical reactions leading to an increase in ultraviolet radiation available to photolyze the N_2O, thereby shortening its atmospheric lifetime. Such feedbacks are accounted for in the above table.

2 The GWP provides a simple measure of the radiative effects of emissions of various greenhouse gases, integrated over a specified time horizon, relative to CO2 emissions. Unless otherwise indicated, GWPs taken from: IPCC (2001).

3 Following the convention of IPCC (2001), inferred global-scale trace-gas concentrations from prior to 1750 are assumed to be practically uninfluenced by human activities such as increasingly specialized agriculture, land clearing, and combustion of fossil fuels.

4 For most gases, concentrations for year 2004 are given, as indicated more specifically in the footnotes below. Estimates for 1998, from IPCC (2001), are given for C_2F_6. Atmospheric concentrations of some of these gases are not constant throughout the year. Global annual arithmetic averages are given.

5 The value given by IPCC (2001), page 185, is 280 ± 10 ppm. This is supported by measurements of CO_2 in old, confined, and reasonably well-dated air. Such air is found in bubbles trapped in annual layers of ice in Antarctica, in sealed brass buttons on old uniforms, airtight bottles of wine of known vintage, etc. Additional support comes from well-dated carbon-isotope signatures, for example, in annual treerings. Estimates of "pre-industrial" CO_2 can also be obtained by first calculating the ratio of the recent atmospheric CO_2 increases to recent fossil-fuel use, and using past records of fossil-fuel use to extrapolate past atmospheric CO_2 concentrations on an annual basis. Estimates of "pre-industrial" CO_2 concentrations obtained in this way are higher than those obtained by more direct measurements; this is believed to be because the effects of widespread land clearing are not accounted for.

6 Recent CO_2 concentration (377.3 ppm) is the average of the 2004 annual values at Barrow, Alaska; Mauna Loa, Hawaii; American Samoa; and the South Pole (one high-latitude and one low-latitude station from each hemisphere).

7 Pre-industrial concentrations of CH_4 are evident in the "1000-year" ice-core records. However, those values need to be multiplied by a scaling factor of 1.0119 to make them compatible with the AGAGE measurements of current methane concentrations, which have already been adjusted to the Tohoku University scale. Therefore, pre-industrial values calculated from the ice-core data have been multiplied by 1.0119 before insertion in the above table.

8 The first value represents Mace Head, Ireland, a mid-latitude Northern-Hemisphere site, and the second value represents Cape Grim, Tasmania, a mid-latitude Southern-Hemisphere site. For CH_4, these values can be compared with the thousand-year ice-core records from Greenland and Antarctica, respectively, discussed in the preceding footnote. "Current" values given for these gases are annual arithmetic averages based on monthly nonpollution concentrations for year 2004.

9 Source: IPCC (2001). The pre-1750 value for N_2O is consistent with ice-core records in IPCC (2001). Estimates of "current" (1998) concentrations of C_2F_6 are based on a variety of sources, including emissions rates and annual growth rates.

Table 2. Main Greenhouse Gases. *Table courtesy of Pew Center on Global Climate Change. Source: Blasing, T.J. and K. Smith 2006. "Recent Greenhouse Gas Concentrations." In* Trends: A Compendium of Data on Global Change. *Carbon Dioxide Information Analysis Center, Oak Ridge National Laboratory, U.S. Department of Energy, Oak Ridge, TN, USA, http://cdiac.ornl.gov/pns/current_ghg.html.*

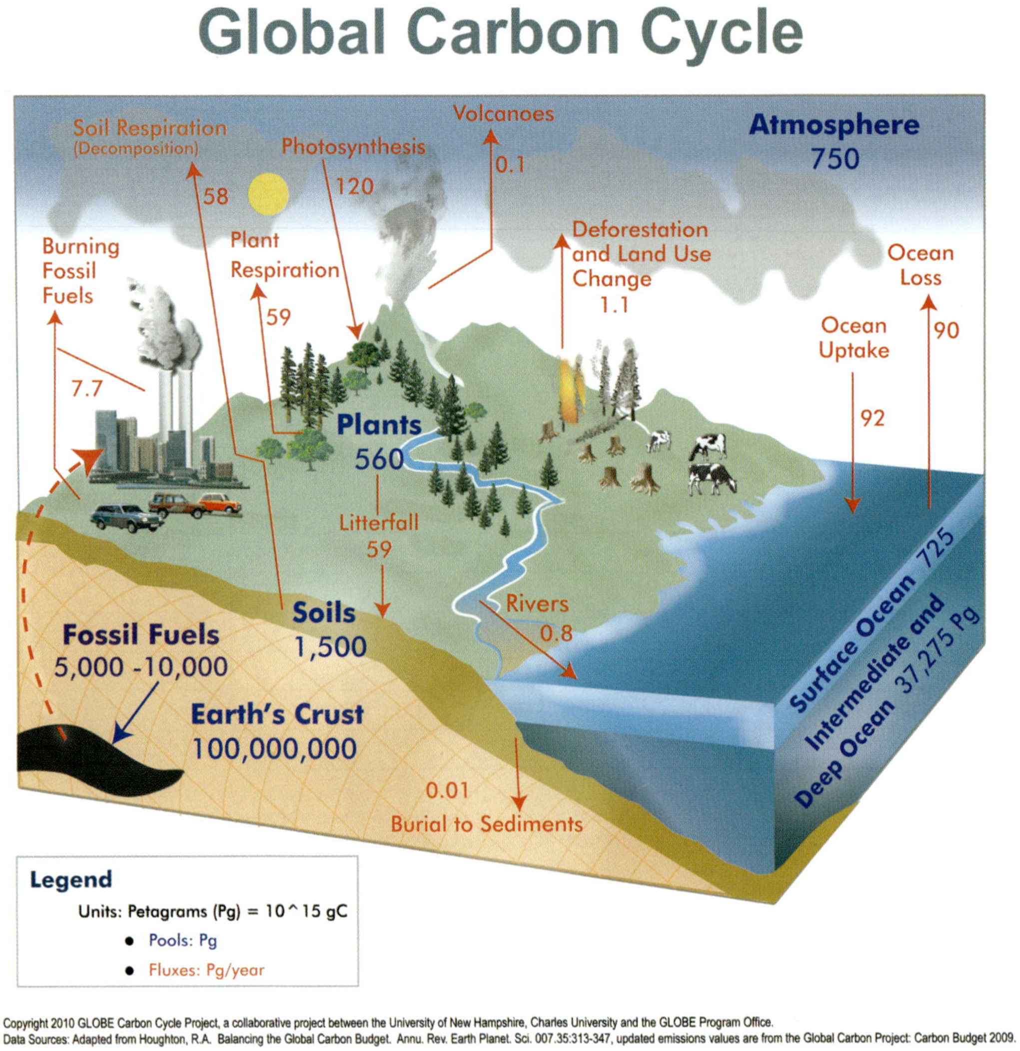

Figure 11. *Sources and sinks of carbon on Earth. The sinks are highlighted in blue and measured in petagrams (1 Pg = 10^12 grams) of carbon. Changes in carbon are noted in red and measured in petagrams/yr (Pg/yr). Source: http://globecarboncycle.unh.edu/diagram.shtml.*

are measured in gigatonnes of carbon (GtC), rather than carbon dioxide.

The process that exchanges carbon dioxide between the atmosphere and the ocean surface involves the dissolving and undissolving of carbon dioxide. Consider a beverage like club soda: in order for the club soda to be "fizzy," carbon is dissolved in water to create carbonated water, a key ingredient in club soda. However, if a bottle of club soda is left open for a period of time, the carbon undissolves, and it is released from the club soda into the air, leaving it "flat."

In much the same way, carbon dioxide in the atmosphere is constantly being dissolved in water on the surface of the oceans, while the sea surface is constantly releasing carbon dioxide back into the atmosphere. The sea surface releases an estimated 88 GtC/year and takes in an estimated 90 GtC/year. As a result, this activity leads to a net loss of 2 GtC/year from the atmosphere, so the sea surface exchange is actually a "sink" for atmospheric carbon dioxide—i.e., it takes out more than it puts back. Some recent evidence suggests that the oceans may be taking up less carbon dioxide than they have in the past, raising concern that atmospheric concentrations could rise faster. Much research is needed to confirm this observation, and confidence levels are low.

The magnitude of these carbon flows relative to other processes is particularly important, because small changes in the delicate balance (or in our estimates of them) could have a large impact.

As important as the physical and chemical processes just described are the biological processes that cycle carbon dioxide to and from the atmosphere. Plants "breathe in" carbon dioxide through the process of photosynthesis—about 120 GtC every year. But plants, animals, and other organisms also "breathe out" carbon dioxide. For example, humans and other land mammals breathe in oxygen, which they burn in the metabolic processes known as respiration. Carbon dioxide is exhaled as a waste product. Together, all the living things are estimated to "exhale" (produce by respiration) about 60 GtC/year.

When plants and animals die, the organic carbon compounds they have stored become part of the soil, forest litter, or swamp muck. Nature composts this **detritus** of life much as gardeners do, by breaking it down through many processes of chemical decomposition and microbial action. As they break down, soils and detritus are estimated to release about 55 GtC/year back into the atmosphere.

So the 120 GtC taken out of the atmosphere every year by plants is almost balanced by the total 115 GtC put back into the atmosphere each year by respiration and decay. Another 4 GtC is estimated to be put back into the atmosphere every year by combustion of biomass—such as wildfire and agricultural burning. The magnitude of this cyclic flow of terrestrial carbon is significant, because small disturbances in the balance can have large implications for the net flow since the difference between sources and sinks is very small.

Is the terrestrial biosphere adding to or subtracting from atmospheric carbon dioxide, when all sources and sinks are netted out over the whole planet for an average year? It is not an easy question and the estimates can be imprecise.

The AR4 estimated that the terrestrial biosphere took up 0.3 GtC/year in the 1980s on an overall net basis, but the margin of error is 0.9 GtC/year, leaving the net impact unclear. For the 1990s, the estimate is 1.0 GtC/year, with a smaller error of 0.6 GtC/year. The AR4 also estimated the net terrestrial uptake for the period 2000–2005 at 0.9 GtC/year with a similar error as in the 1990s.

Compared with the huge amounts of carbon the atmosphere exchanges with the ocean and land ecosystems, the amount of carbon dioxide added directly to the atmosphere as a result of human activities might at first seem inconsequential. By burning coal, oil, and natural gas—withdrawing from nature's bank account, as it were—society currently takes an estimated 7.2 GtC/year out of the Earth and puts it into the atmosphere, based on many different peer-reviewed journal articles and reported in the AR4.

But these amounts do matter, because the natural parts of the carbon cycle (the air-sea exchange and the biological processes) have long been in close balance, at least on the timescales of immediate relevance to humans. Human industrial activities, agriculture, and land-use changes seem to have significantly tipped the balance of the carbon cycle.

Many kinds of scientific measurements have shown that the concentration of carbon dioxide in the atmosphere has been increasing over the past several centuries. During this time the human population increased geometrically, the steam engine was put to industrial use, the gasoline-powered automobile came into use across the globe, and farmer-settlers cleared native vegetation from vast expanses of the Americas, Australia, and parts of Asia.

During this same period, the atmospheric concentration of carbon dioxide has increased from a pre-industrial (1750) level of about 280 parts per million by volume (ppmv) to about 379 ppmv in 2005, an increase of more than a third. Far removed from the potential influence of local or regional industrial pollution sources, precise measurements taken at the Mauna Loa Observatory in Hawaii have charted steady increases from 1958 to 2001 (Figure 12), with a brief pause between 1990 and 1992.

The close relationship between carbon dioxide concentrations and estimated global mean temperature over longer periods in the Earth's history is striking. Figure 13a shows that during cold periods, CO_2 concentrations are low, and during warmer periods those concentrations are higher. It is tempting to conclude that fluctuations in CO_2 caused most of the observed temperature changes. Whether the relationship is a causal one, how-

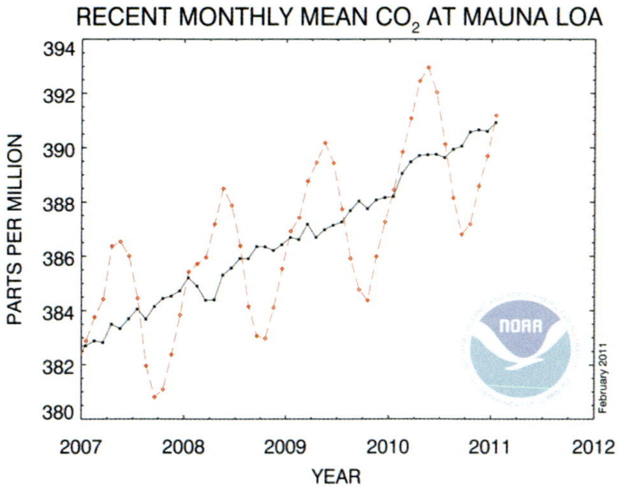

Figure 12. *This figure shows the monthly mean value of CO_2 concentrations as measured from the top of Mauna Loa in Hawaii since 2007. The red curve indicates how the concentration levels change with the seasons: as the Northern Hemisphere greens in the spring and summer, the vegetation takes up CO_2, reducing the amount in the atmosphere. The black line shows the same data with the seasonal effect removed, clearly illustrating the upward trend over time. Source: NOAA, http://www.esrl.noaa.gov/gmd/ccgg/trends/.*

ever, is unclear, because the causality seems to work both ways. The colder temperatures on Earth during glacial periods (ice ages caused by the Earth's orbit rather than greenhouse gases) seem to lower CO_2 concentrations, not by reducing plant life as might be expected, but by changing the oceans in ways not yet understood. Likewise, warmer temperatures elevate CO_2. One needs to look at external drivers of past climate change to evaluate which causes which. However, in the 20th and now 21st centuries it is clear that human activities are driving changes in the carbon cycle that are leading to changes in climate.

In the 20th and now 21st centuries it is clear that human activities are driving changes in the carbon cycle that are leading to changes in climate.

Figure 13 demonstrates that CO_2 and temperature are tightly linked by a number of complex feedback mechanisms. The bad news is that such feedbacks could amplify the effects of greenhouse warming caused by human emissions.

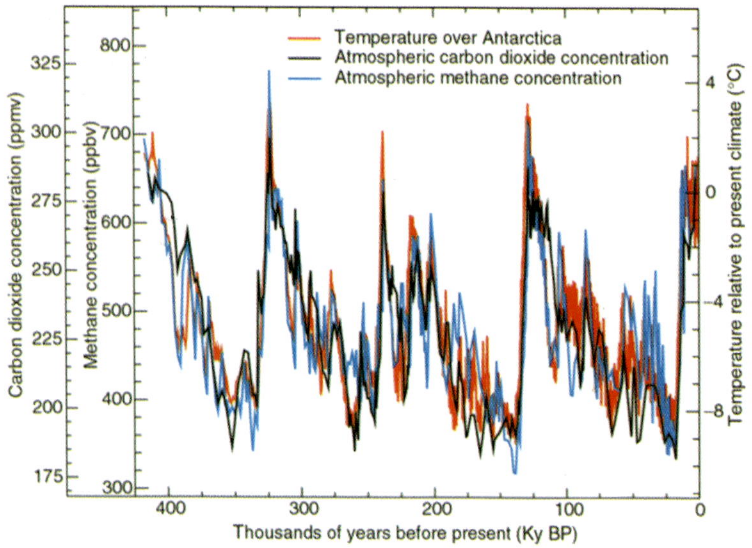

Figure 13. *Variations of temperature, methane, and atmospheric CO_2 concentrations presented here are derived from air trapped within ice cores from Antarctica. Source: IPCC, Climate Change 2001: The Scientific Basis (adapted from Sowers and Bender, 1995; Blunier et al., 1997; Fischer et al., 1999; Petit et al., 1999).*

METHANE

There are many natural and human-related sources of methane in the atmosphere. Methane (CH_4) is the major constituent of what comes out of the burner on most gas stoves—natural gas. As a trace gas in the atmosphere, methane has a greater warming effect than carbon dioxide—25 times more, pound for pound

Concentrations of methane in the atmosphere have more than doubled since the Industrial Age began—about eight times faster than carbon dioxide. The concentration rose from an estimated 700 parts per billion by volume (ppbv) in the year 1750 to 1,774 ppbv in 2005. Growth rates in methane concentrations slowed beginning in the early 1990s and were approximately zero from 1999 to 2005, meaning that emissions and removals were about equal. That growth rate has also changed greatly from year to year, which is not fully understood.

Some good news on methane as a greenhouse gas: its atmospheric lifetime is much shorter than that of most other greenhouse gases—probably about 12 years. Most methane is removed from the atmosphere when it combines with the hydroxyl radical (OH) to form water and carbon dioxide.

It remains unclear how much methane comes from natural sources and how much from human sources. Approximately 582 teragrams of methane ($TgCH_4$) are released into the atmosphere each year from all sources, both natural and anthropogenic. The totals in the following discussions are tentative estimates.

Methane is one of the gases formed in the Earth's interior, and it is vented through volcanoes and other breaks in Earth's crust. The amount of methane released annually from natural geologic sources has not been measured, although it is not believed to be important.

Much atmospheric methane is biological in origin. Methane is produced by bacteria in the absence of oxygen (the **anaerobic** bacteria). These bacteria decompose the plant and animal refuse in the black muck of natural wetlands, such as swamps and marshes.

Another name for methane is "swamp gas" (its characteristic odor comes mostly from the hydrogen sulfide mixed with it). Wetlands are estimated to produce 100 to 245 $TgCH_4$/year, another major fraction of the global total.

Rice paddies, which are agricultural wetlands specifically created for cultivating rice, are estimated to contribute another 31 to 112 $TgCH_4$/year.

Another important source of methane is intestinal gas—especially from cattle, water buffalo, sheep, and other ruminant livestock that humans raise in farming. Anaerobic microorganisms in the guts of these animals make digestion itself possible—but the methane they produce is vented to the atmosphere. This **enteric fermentation** from all animals is estimated to produce about 80–92 $TgCH_4$/year, an appreciable amount. Intestinal gas is a natural source of methane, but the amount of this "natural source" varies based on human activity. Increases in the amount of human-raised livestock result in an increase of methane released.

Among the main anthropogenic sources of methane are the byproducts of energy extraction industries—coalbed methane (30 to 46 $TgCH_4$/yr), and various emissions from gas pipelines, oil wells, and industry (52 to 68 $TgCH_4$/yr). Other anthropogenic methane sources include agricultural animal wastes, sewage treatment, and landfills (35 to 69 $TgCH_4$/yr), and biomass burning (14 to 88 $TgCH_4$/yr).

These estimates have changed significantly in the past few decades and remain quite uncertain. Whatever the uncertainties of the estimates, it seems likely that much of the observed increase in atmospheric methane concentrations since the industrial age began has been caused by human activity. It follows that, without technological change, further increases in human population, industry, and agriculture will bring further increases in atmospheric methane concentrations.

NITROUS OXIDE

Nitrous oxide (N_2O) is commonly known as "laughing gas." Its building blocks are the two dominant gases in the atmosphere: nitrogen and oxygen. Both play important roles in the chemistry of living things (nitrogen goes into amino acids, the "building blocks" of protein and key to metabolism, growth, and repair of tissue). Although nitrogen gas (N_2) and oxygen gas (O_2) make up some 99% of the atmosphere, nitrous oxide is scarce—with a mean concentration of 319 ppbv (parts per billion by volume) in 2005, according to the AR4 (i.e., less than one thousandth as abundant as carbon dioxide).

The scarcity of nitrous oxide in the atmosphere is countered by its effectiveness as a greenhouse gas. Because of differences in the wavelengths it absorbs, nitrous oxide is almost 300 times more effective than carbon dioxide (on a per mass basis) in the greenhouse warming it produces. Furthermore, it is longer lived in the atmosphere than carbon dioxide. Nitrous oxide's atmospheric lifetime is about 114 years. It is eventually broken down by ultraviolet light in the stratosphere into nitrogen and oxygen.

Microorganisms, alone or in symbiosis with plants, are constantly taking nitrogen out of the atmosphere and putting it into the soil. Nitrogen in the atmosphere (N_2) is fairly inert and largely unusable by living things. The triple bond between the two nitrogen atoms makes it hard for them to recombine with other atoms to make new compounds. Microorganisms convert N_2 into other forms, such as the ammonia ion (NH_3) or the nitrate ion (NO_3), which can be used by plants to make other chemical compounds because the bonds holding the nitrogen atoms are weaker. This conversion process is called **nitrogen fixation**. After nitrogen is fixed, it may be repeatedly cycled from the soil into plants and animals.

All the while, other microorganisms, in a process called **denitrification**, are constantly taking nitrogen out of its fixed form in the soil and putting it back into the atmosphere. Besides yielding molecular nitrogen, denitrification produces nitrous oxide. Scientists estimate that soil denitrification is the dominant source of atmospheric nitrous oxide. Another important contribution comes from natural ocean processes, which are not well understood.

At the beginning of the industrial age, atmospheric concentrations of nitrous oxide were steady at about 270 ppbv, so the current concentration of 319 ppbv represents an 18% increase. This increase is believed to result from human activities. Combustion is one source of nitrous oxide (burning fossil fuels, wood, or other biomass). But scientists today think combustion may be less important a source of anthropogenic nitrous oxide than agricultural soils and the use of fertilizer.

Worldwide, fertilizer use puts some 55 teragrams (Tg, or 1,012 grams) of nitrogen into the soil each year. Nitrogen fertilizer is made either by mining nitrates or by fixing atmospheric nitrogen (into the usable form of nitrate or ammonium) by industrial processes. When this artificially enriched soil is denitrified, or when fertilizers leach into groundwater, nitrous oxide goes into the atmosphere.

Estimates of the amount of nitrous oxide put into the atmosphere vary widely. According to the latest mid-range estimates cited by the IPCC, natural sources such as oceans, atmosphere, and soils account for roughly two-thirds of the N_2O going into the atmosphere, with the remaining one-third coming from human activities. Agricultural soils account for about 40% of the total human contribution. Most of the remaining human contribution comes from cattle and feedlots; industrial sources like production of nylon and nitric acid or fossil-fueled power plants; or biomass burning.

HALOCARBONS

Also intensifying the greenhouse effect in the Earth's atmosphere are halocarbons. These compounds combine carbon with one or more of the five elements called **halogens**: fluorine, chlorine, bromine, iodine, and astatine, but only the first three are relevant here. Included in the family of halocarbons are the chlorofluorocarbons (CFCs); the hydrochlorofluorocarbons (HCFCs); HCFC substitutes; and some others—carbon tetrachloride, halons, methyl chloride, methylchloroform, and methyl bromide.

Some of these gases, such as CFCs -11, -12, -113, -114, and -115, have gotten quite a bit of press attention. They typically were used in the United States until the mid-1970s in spray-can propellants, and as solvents, cleaners, and coolants. Many nations agreed to control emissions of these chemicals in 1987 when they signed the Montreal Protocol on Substances that Deplete the Ozone Layer. (See Chapter 8, Stratospheric Ozone.)

The other role halocarbons play—as greenhouse gases—is often overlooked but is also important.

All of the halocarbons in the atmosphere result from human activities, except for methyl chloride and methyl bromide, which also have important natural sources. The concentration of methyl chloride does not appear to be growing, but the concentration of methyl bromide is. The concentrations of halocarbons in the atmosphere are much lower than those of the other greenhouse gases, typically between 0.2 and 16.5 parts per trillion by volume.

Halocarbons generally raise some concern because their greenhouse warming effect, on a molecule-for-molecule basis, is substantially greater than that of carbon dioxide. The five ozone-depleting CFCs of most concern have warming effects ranging from 3,000 to 13,000 times greater than carbon dioxide.

Halocarbons also raise concerns because of their longevity. Chemical engineers invented some of these chemicals specifically for their stability, a valuable quality given their industrial purposes. But once in the atmosphere, that very stability becomes a problem: they resist breakdown and removal for many decades. The atmospheric lifetimes of CFC-13 and CFC-115, among the longest-lived, are about 400 years, so their harmful effects will continue for centuries even after they no longer are released into the atmosphere.

Reductions and elimination of CFC production under the Montreal Protocol will, as a beneficial side effect, help slow the growing concentrations of these greenhouse gases. But some halocarbons with greenhouse potential are not restricted under that international agreement, and some of the CFC substitutes whose uses it encourages are themselves halocarbons with greenhouse effects.

TROPOSPHERIC AND STRATOSPHERIC OZONE

News reports have cast ozone in the roles of both villain and victim over the past decade. Ozone plays a number of complex, everyday roles in the atmosphere, and is constantly being created and destroyed.

Ozone makes the sweetish-pungent smell coming from the arc of electric motors and some office copier machines. It is also one of the principal components of the urban smog that irritates the eyes and lungs.

We are likely to have ozone around as long as we have an atmosphere with abundant oxygen. In its ordinary form as a free element, oxygen usually occurs as a diatomic molecule (consisting of two oxygen atoms)—O_2. Ozone is a triatomic molecule consisting of three oxygen atoms—O_3—formed when certain kinds of energy, say an electric arc or very intense ultraviolet light, are applied to O_2.

Most of the ozone in the atmosphere is created when the high-energy ultraviolet rays of the Sun reach the rarefied gas molecules of the stratosphere. The ultraviolet energy excites the oxygen atoms, and they combine to form ozone. But ultraviolet rays not only create ozone, but also destroy it when lower-energy ultraviolet breaks apart the ozone molecules.

Most of the naturally occurring ozone tends to be concentrated in the lower stratosphere, but some ozone molecules do find their way down into the troposphere, accounting for about half of the ozone there. The other half is produced in the troposphere both from natural sources and as a result of human activities. Humans add some ozone in the troposphere through various kinds of air pollution. When automobiles and other human sources emit carbon monoxide, methane, and nonmethane hydrocarbons in the presence of nitrogen oxides, sunlight causes a reaction that produces ozone "smog," which raises important public health issues in many major urban areas.

Ozone in the upper troposphere and lower stratosphere functions as a greenhouse gas. Just how strong its greenhouse effect is, compared to carbon dioxide, is something scientists are still trying to quantify. But ozone's effect seems to be weaker—with an average forcing estimated between 0.25 and 0.65 Watts per square meter (W/m^2, a standard measure of the strength of a gas's greenhouse effect in terms of energy over a given area of the earth's atmosphere), compared to 1.66 W/m^2 for CO_2. Measurements of ozone concentrations, especially of historical trends, are few.

Ozone is unique among greenhouse gases in the brevity of its atmospheric lifetime. Because it is dynamically created and destroyed, its atmospheric lifetime is measured in weeks. Thus, ozone and its greenhouse influence cannot be expected to accumulate in the atmosphere, as is the case for other greenhouse gases. However, the greenhouse implications of ozone are complex, because of its interactions with the halocarbons when it is transported to the upper troposphere (with increases in one greenhouse gas causing reduction of another). In addition, the warmer and more humid weather that climate change is expected to bring to many regions dramatically enhances the formation of ground-level ozone.

WATER VAPOR

The water vapor in the atmosphere is essentially molecules of steam—evaporated water—thinly diluted and bouncing around among the gaseous nitrogen and oxygen molecules that make up the atmosphere. This water vapor is not well mixed into the atmosphere like the other gases, and its concentration varies greatly across the globe, with the greatest concentrations over the tropical oceans where the most evaporation from the surface occurs. Water vapor can be experienced as humidity (hence the humid tropics).

Although water vapor is the third most abundant component in the Earth's atmosphere, and it has a greenhouse effect, it is not a driver (**forcing**) of climate change; rather, it acts as a **feedback**, amplifying the warming effect caused by the increase in the six main GHGs. (See Forcing vs. Feedback, Chapter 2).

Water vapor plays a number of critical roles in affecting both climate and weather. The amount of water vapor in the atmosphere is not at all uniform, far from it: it changes drastically and abruptly, often in a matter of a few hours, to cause, for example, thunderstorms.

It takes a lot of energy to evaporate liquid water into water vapor. Therefore, a molecule of water vapor "contains" much more energy than a molecule of liquid water. Since quite a bit of water is evaporated everyday as the Sun shines on the Earth's vast oceans, water vapor is one of the most important storehouses and transport mechanisms of heat in the atmosphere and in the climate system.

Since water vapor has more energy than liquid water, it loses energy when it condenses into the tiny suspended droplets that make clouds, or into the larger drops constituting rain. The "lost" energy does not disappear, but instead heats the atmosphere. Thus, energy is redistributed through the processes of evaporation and condensation.

When water vapor condenses to form clouds, it has another important effect: it "shades" the Earth's surface and lower atmosphere. In the greenhouse analogy, rolling down a shade over the greenhouse would cool off the interior, just as it would cool a sunny room. (Cloud formation is an important process in the climate system, but one that is hard to quantify and model.) When clouds shade the Earth, some of the incoming solar energy is reflected back into space. Some also is absorbed by the clouds and reradiated upward and downward. Thus, some of the solar energy is caught at altitudes higher than the Earth's surface, but still in the atmosphere.

Although water vapor can condense and have a cooling effect, in its vapor state it also has an important heat-trapping greenhouse effect. Like the other GHGs, water vapor is relatively transparent to the shorter wavelengths of the electromagnetic (or visible and ultraviolet) spectrum, but much less transparent to the longer infrared wavelengths. With its short wavelength, incoming solar energy easily passes through the water vapor. However, after this energy has warmed the Earth's surface and is reradiated upward as infrared light, water vapor readily absorbs it—trapping heat in the lower atmosphere (troposphere). Thus, the water vapor is like the heat-trapping "glass" in the greenhouse analogy.

Eventually, this trapped heat finds its way upward and outward and is reradiated into space. But first it works its way through various parts of the atmosphere. Because incoming solar radiation (now outgoing heat radiation) is thus delayed in returning to space, the temperature of the lower atmosphere is greater than it would be without the water vapor.

Human activities add and subtract water vapor to and from the atmosphere, but these amounts are insignificant compared to the amounts added and subtracted by natural processes. Today's climate models are correctly accounting for the effects of water vapor. In fact, the amount of water vapor in the atmosphere is determined by the climate at the same time that it strongly affects the climate—a classic **feedback loop**. Water vapor is typically not listed with other changing greenhouse gases primarily because the science community understands that changes in water vapor are a feedback response to changes in climate, and not a force external to the climate system (as are human changes to other greenhouse gases)—see box. For example, warmer air holds more moisture, which means that as atmospheric temperatures rise from the increase in GHGs, the air contains more water vapor. The added water vapor amplifies the warming.

As discussed later, in Chapter 10, there is substantial scientific uncertainty about water's role as vapor or clouds. The ultimate effect of water vapor on climate is complicated by its influence on cloud formation. The AR4 estimates that the expected increase in atmospheric water vapor due to warming temperatures will cause approximately a 50% amplification of global average warming.

Observed Changes and Trends in Greenhouse Gas Concentrations

Along with a strong scientific consensus on the basic physics governing the greenhouse effect, as discussed above, scientists are sure that there has been an increase in the atmospheric concentration of greenhouse gases, and especially of carbon dioxide, over recent decades and centuries. The increase is measurable, both directly and indirectly.

Precise measurements of atmospheric CO_2 have been taken consistently from instruments at the Mauna Loa Observatory atop a mountain in Hawaii, where the atmosphere is relatively undisturbed by immediate pollution from the urban areas on the continents (Figure 11). The air samples from Mauna Loa are well mixed, averaging out any local effects of human pollution on the atmosphere at large.

Those measurements show a steady upward trend in the CO_2 concentration since 1958 when measurements began. It increased from about 315 ppmv in 1958 to about 379 ppmv in 2005, according to the AR4. Additionally, scientists can get historical air samples by collecting tiny bubbles trapped in the thick ice sheets of Antarctica and Greenland. Measurements of these samples show with little doubt that the CO_2 concentration has increased from the pre-industrial value (~1750) of about 280 ppmv.

Are these increases the result of activities undertaken by people? The vast majority of scientists studying the matter are convinced that most of the increase results from human actions. First, the increase is much larger than the natural variability they observe in CO_2 concentrations over thousands of years. Second, they know how much coal and oil industrial-age societies have burned, and how much forest they have cut down for agriculture, and these factors are roughly double the observed increase, which balances with the observed carbon sinks. Third, isotope analysis of the carbon in atmospheric CO_2 suggests that much of the increase did come from fossil fuel burning. Fourth, complex models of the carbon cycle that represent the important processes and feedbacks between the atmosphere, biosphere, and oceans cannot explain the observed changes in CO_2 without the human component. See Chapter 9 for a discussion of how scientists detect and attribute changes in the Earth's climate.

CHAPTER 4: CLIMATE CHANGE IS HAPPENING NOW

Assessing the current state of the science of climate change, the AR4 solidified the scientific consensus, which is stronger now than in any previous report. It concluded that the increase in global average temperature is *beyond doubt*: "Warming of the climate system is unequivocal, as is now evident from observations of increases in global average air and ocean temperatures, widespread melting of snow and ice and rising global average sea level." (See Figure 1 in Chapter 1).

It is worth emphasizing that scientists rarely reach such unambiguous conclusions. That the vast majority of scientists (as well as some 120 of the world's governments, many of which are concerned about the potential costs of stopping climate change) were able to agree on this language reflects the strength of this conclusion. *No honest broker of information can argue that the Earth has not warmed, because there is simply too much evidence of warming and no meaningful evidence to the contrary.*

Having said that, the next logical question is, Are humans to blame? Here the AR4 made a much stronger statement than any previous assessment, stating:

> Most of the global warming in the past 50 years is *very likely* [>90% probability] due to human increases in greenhouse gases. (emphasis added)

Compared to previous assessments, this represents a strengthening of the conclusion that humans are playing the dominant role in driving changes in the Earth's climate. But how do scientists know this? They base the conclusion on two basic principles:

(1) Detection: Has warming been observed?

(2) Attribution: If so, what is causing it?

Detection

In order to know whether the world is getting warmer, scientists need to have pretty good data on past and present temperatures. Fortunately, the satellite era offers a wealth of information on temperature as well as a number of other climate variables. Combined with surface thermometers (land and sea) dating back to 1850, scientists have developed a robust record of temperature, and by studying these data they have reached the following conclusions:

- According to the latest analysis by the National Oceanic and Atmospheric Administration (NOAA), *nine* of the last ten years have ranked as the hottest on record globally. NOAA's analysis indicates that 2010 is tied with 2005 as *the* hottest year on record.
- Anthropogenic warming of the climate system is widespread and can be detected in temperature observations taken at the surface, in the free atmosphere, and in the oceans.

In addition, changes in global temperature affect precipitation patterns, and these effects have also been observed in the recent climate record:

- Long-term trends from 1900 to 2005 have been observed in increased precipitation amount in many large regions.
- Increases have occurred in the number of heavy precipitation events.

Attribution

Attribution is the process of establishing cause and effect with some defined level of confidence, including the assessment of competing hypotheses. Demonstrating cause and effect is not always easy. In science, the preferred way of demonstrating cause and effect is the experimental method—devising experiments specifi-

> **THE BOTTOM LINE ON OBSERVED CLIMATE CHANGE**
>
> - Warming of the climate system is unequivocal—there is literally *no doubt* that an increase in global average temperature has been observed.
> - Most of the warming observed in the past 50 years is, with >90% certainty, due to human activities that release greenhouse gases into the atmosphere.
> - The evidence for the conclusion that increases in greenhouse gases are the primary cause of recent warming and associated changes in climate has continued to build over the last two decades, dramatically in the past five years.

cally to test a causal hypothesis, and then observing the results. The problem, of course, is that the Earth is not something we can experiment with in the ordinary sense. Some would argue, of course, that we *are* experimenting with the Earth—that we are changing a key variable and may soon observe a result.

Even though scientists cannot conduct experiments on the Earth in the traditional sense, there are many observations that scientists can make that will help establish the cause of the current warming. What scientists know about physical climate processes (as extended through models) suggests that warming forced by anthropogenic greenhouse gas increases will take place in certain specific ways. They can then observe whether the warming is happening in these ways. These attribution studies are sometimes called **fingerprinting**.

Scientists need to know enough about Earth's climate history to understand the planet's natural climate variability, in order to determine whether observed changes could be accounted for based on natural variance alone. (See Chapter 2, Natural Variability.) Natural variability consists of two things—changes in natural forcings (e.g., volcanoes, orbital shifts, solar variability) and internal climate system variability. Internal variability, we have learned, includes certain natural cycles (like El Niño) and also random, or chaotic static.

Three key insights lead scientists to attribute the detected warming trend to the GHGs released by human activities.

(1) Conservation of energy dictates that simultaneous warming of the atmosphere and oceans can only occur due to an external forcing. This observed pattern of change rules out internal variability.

(2) The vertical pattern of atmospheric temperature change shows warming in the troposphere (lower atmosphere) and cooling in the stratosphere; this pattern carries the GHG fingerprint of external forcing.

(3) Physical matching between the temporal and spatial patterns of the warming trend and changes in multiple radiative forcings also reveals the GHG fingerprint for external forcing.

What does the AR4 have to say about attribution of recent warming to human activities?

- Greenhouse gas forcing has *very likely* [>90%] caused most of the observed global warming over the last 50 years.
- It is *extremely unlikely* (<5%) that the global pattern of warming during the past half century can be explained without external forcing, and *very unlikely* [<10%] that it is due to known natural external causes alone. The warming occurred in both the ocean and the atmosphere, and took place at a time when natural external forcing factors would *likely* [>66%] have produced cooling. See Table 5, Appendix A.

The job of attribution is made easier by the fact that competing hypotheses are relatively few—and relatively weak. None of these competing hypotheses—increased solar irradiance, decreased volcanism, natural oscillations (or internal variability), etc.—adequately explains

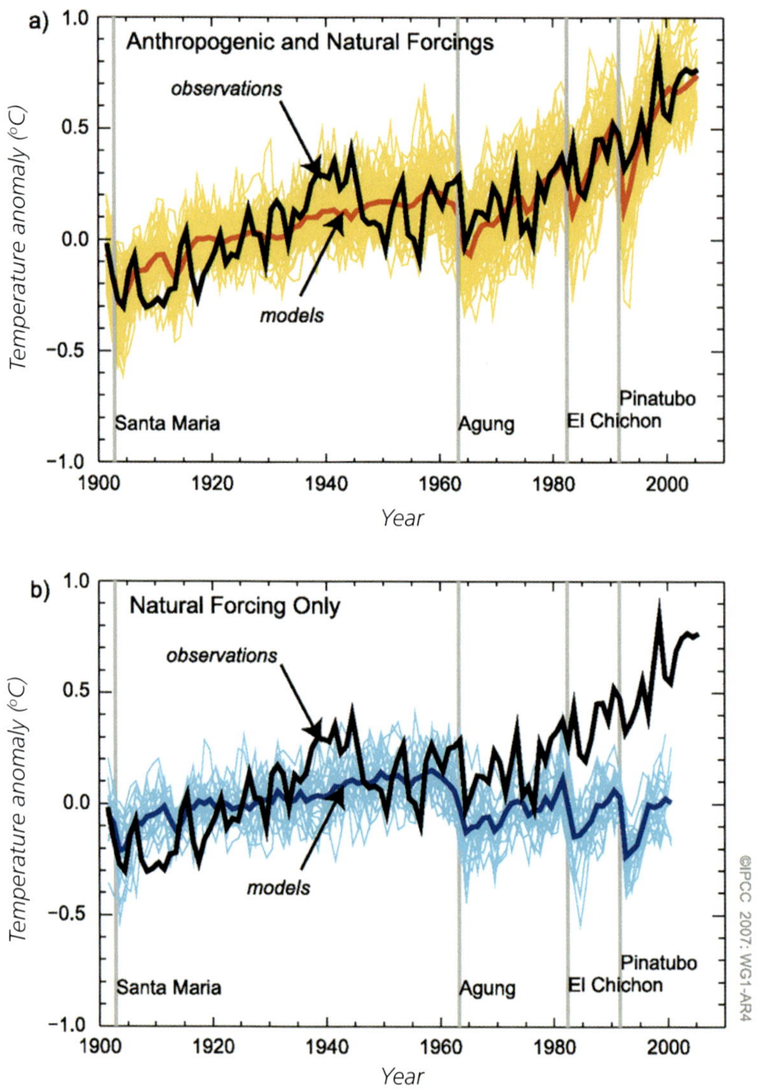

Figure 14. *Shown here is the temperature anomaly—the difference between observations and the average from the period 1901–1950. Observations are shown by the black line. Model simulations in the upper plot include both natural and human-induced effects, while the lower plot includes natural effects only. It is evident from these plots that human effects must be included in order to match the observed data. Source: IPCC, AR4 WGI, Technical Summary, Figure TS.23.*

the observed data. They would not be strong enough to account for warming as great as scientists have observed. In addition, any competing hypothesis would also have to explain why increased concentrations of greenhouse gases would *not* increase temperatures, which basic physics tells us is the case.

In fact, research published since the AR4 has only strengthened the connection between climate change and human activities. Other fingerprinting studies have linked human-induced warming to climate effects such as increases in polar temperatures in the Arctic and Antarctic, global precipitation trends, changes in the hydrological cycle in the western United States, an increase in total atmospheric moisture over the oceans, an increase in sea surface temperatures in regions of hurricane formation in both the Atlantic and Pacific Oceans, and various physical and biological changes, such as the timing of seasonal lake freezing and thawing and the timing of plant flowering and animal migration. These studies are by no means the final word on fingerprinting, and they each contain uncertainties, but again, multiple lines of observational evidence point to humans as the dominant driver of climate change we are already observing.

> *I think that much of the foot-dragging in addressing climate change is a reflection of the perception that climate change is way down the road . . . and that it only affects remote parts of the planet. And this report demonstrates . . . that climate change is happening now and it's happening in our own backyards and it affects the kinds of things people care about.*
>
> —Dr. Jane Lubchenco, NOAA Administrator
> GCRP Press Conference, June 2009

Increased Confidence

To illustrate the increased scientific consensus on the human contribution to global climate change, we can look at the statements issued by the IPCC in its last three assessment reports:

IPCC 1st Assessment Report (FAR): 1990

> The unequivocal detection of the enhanced greenhouse effect from observations is *not likely* for a decade or more. (emphasis added)

IPCC 2nd Assessment Report (SAR): 1995

> The *balance of evidence suggests* a discernible human influence on global climate. (emphasis added)

IPCC 3rd Assessment Report (TAR): 2001

> Most of the observed warming over the last 50 years is *likely* [>66% probability] to have been due to the increase in greenhouse gases. (emphasis added)

IPCC 4th Assessment Report (AR4): 2007

> Most of the global warming in the past 50 years is *very likely* [>90% probability] due to human increases in greenhouse gases. (emphasis added)

These statements demonstrate a change in the scientific consensus over time toward increasing confidence in the existence of human-induced warming. The evidence for this conclusion has continued to build over the last two decades.

CHAPTER 5: THE HUMAN EFFECT ON CLIMATE

The idea that humans could change the planet Earth is a relatively new one in human history. For millennia, people felt small and powerless before the awesome size of the Earth and the enormous forces of nature.

Figure 15. *This composite image, which has become a popular poster, shows a global view of Earth at night, compiled from over 400 satellite images. Source: NASA.*

But as agricultural and industrial civilizations arose, and human population began to grow geometrically, people did start changing the face of the planet significantly. We have shaped the vegetation covering the land, the contour and composition of the land, the water flowing over it, the animals living on it, the creatures living in the oceans, and the content of the atmosphere itself. Climate change is just one particular kind of global change caused by humans. Geoscientists have even coined the term "Anthropocene" to describe the postindustrial period.

Climatologists use the term **forcing** to describe phenomena like the human-caused greenhouse warming effect (see box in Chapter 3). The idea is that the climate system is a mechanism (albeit a very complicated one) with an **equilibrium** state (albeit one with many oscillations and internal variabilities). A forcing is any external influence that disturbs or changes the system's equilibrium. The anthropogenic greenhouse effect is a **radiative forcing** on the climate system, as are changes in the Sun's energy. Climate forcings are quantified in terms of energy per unit area—Watts per square meter (W/m^2).

radiative forcing: strength of the forcing

Figure 16. *Burning peat swamp forest in Indonesia, 1997–98. Copyright Dr. F. Siegert. Deforestation causes some 10–20% of the increase in atmospheric carbon dioxide concentrations.*

So, what are the ways in which humans are affecting the Earth's climate?

Deforestation

When Europeans first arrived on North America, the landscape looked a lot different than it does today. Forests are believed to have covered much of the United States east of the Mississippi, and tall-grass prairie, now mostly gone, covered much of the central Great Plains. Those trees and soils contained huge amounts of carbon. As early pioneers and settlers cleared the forest to build farms, trees were cut and burned, and much of the carbon was returned to the atmosphere in the form of carbon dioxide.

The practice of clearing forests to accommodate human settlements and agriculture predates the exploration and settling of North America. Similar processes have occurred over the entire globe during the millennia since human agriculture began, around 9000 BC.

Estimating the net atmospheric effects of various human changes to land cover is difficult for several reasons. First, many of those changes were made in the distant historic or prehistoric past, before precise records were kept. Second, even with today's advanced satellite-mounted instruments and massive databases, land-use change is difficult to measure. Third, the complexities and interconnections of land cover, soils, and other ecosystem components make net emissions or uptake even harder to measure than changes in land cover.

The best available estimates from measured data and modeling suggest that land-use change caused an increase in CO_2 concentrations of between 12 and 35 ppm from 1850 to 2000.

Currently, tropical deforestation appears to be the main factor in land-conversion CO_2 emissions. In the 1990s, about 20% of human-caused emissions of CO_2 were from land-use change—dominated by deforestation. The net emission during the 1980s was estimated at 1.4 GtC/yr (with a range of 0.4 to 2.3 GtC/yr). For the 1990s, it was nearly the same, 1.6 GtC/yr (with a range of 0.5 to 2.7 GtC/yr).

The planting of new forests (**afforestation**) and the replanting or regrowth of forests that have been cut down (**reforestation**) offset some of the deforestation that occurs worldwide. For example, in parts of the Mid-Atlantic region of the U.S. Eastern Seaboard, farmlands once devoted to crop production are becoming covered with second-growth forests, the result both of forestry industry reforesting practices and of natural processes. On the other hand, the loss of tropical rainforests is of particular concern not only because these ecosystems are home to many diverse and unique species, but also because they play a critical role in the carbon cycle.

Change of Other Land Forms

While forest plants and soil are estimated to contain 45% or more of the total carbon stocks in the terrestrial biosphere, other landforms also store significant amounts of carbon.

Grasslands and shrublands in the tropical and temperate regions also contain major amounts of carbon—far more in their soils than in the plants themselves. Humans disrupt these ecosystems in numerous ways: agriculture, planting and exotic plant introduction, grazing, fire (and fire suppression), erosion, etc. The net effect of all these human changes may, by some estimates, result in a net uptake of carbon.

Wetlands, and in particular peatlands (a specific type of wetland), also store major amounts of carbon. Wetland loss—which has occurred as a result of human population growth in the United States, for example—tends to result in loss of carbon to the atmosphere. The anaerobic decomposition typical of wetlands, however, tends to result in the release of the greenhouse gas methane to the atmosphere.

Over the millennia, peatlands have slowly accumulated carbon taken from the atmosphere. These wetlands support vegetation that converts atmospheric CO_2 to biomass, but the accumulating dead biomass decays slowly because of low-oxygen soil conditions and low temperatures. The total carbon stored in peatlands is estimated at some 455 GtC. Humans in various parts of the world disturb, drain, destroy, or mine peatlands. Peat taken from them is eventually burned or decomposes, returning carbon to the atmosphere. (Figure 16.)

Agriculture

Increasing agricultural land is the principal reason humans have cut down forests over the millennia. One of the most primitive methods is often called "slash and burn" agriculture: killed trees and brush are burned onsite, releasing most of the carbon stored in vegetation immediately to the atmosphere. Often a comparable amount of carbon is stored in the decomposing organic matter of the forest soil—and this may be lost within a few years via decomposition and erosion. From a climate perspective, this is more or less a "worst-case scenario."

Soil is in fact one of the major reservoirs of carbon on the planet—that is, carbon that cycles on timescales of immediate relevance to humans. The planet's soils are estimated to contain roughly three times as much carbon as all its plants (trees included). Decaying organic matter (which holds much of the carbon) is one of the key reasons soil is so fertile—which is why farmers put manure on their fields and plow crop residue back under. The soil's carbon content consists of many other things too: vast root systems; worms and grubs; bacteria, fungi, and all manner of other microbes; and carbon in inorganic, mineral forms as well.

However, some farming methods rob the soil of its treasure by either failing to replace the carbon, nitrogen, and other nutrients taken from the soil by plants, or allowing or promoting erosion to remove the topsoil altogether. Wholesale destruction of topsoil has taken place in many parts of the United States throughout its history—from tidewater tobacco lands of the mid-Atlantic and South to the 1930s Dust Bowl in the southern and Great Plains—until the government began to study, teach, and encourage land conservation methods. The best farming methods actually build the soil. They involve high-yielding plant varieties, various fertilizers, proper irrigation, crop residue management, erosion control, and reduced tillage. The soil inputs cost farmers money, but they are an investment in the long-term future of the particular farm. And by banking carbon and nitrogen in the soil rather than in the atmosphere, they also benefit the planet's climate future. Such practices were increasing the net storage of carbon in U.S. agricultural soils by an estimated 0.14 GtC/yr during the 1980s.

The IPCC estimates that improving farming practices worldwide could potentially increase the sequestration of carbon by 0.4 to 0.9 GtC/yr, for a cumulative 24 to 43 GtC over 50 years; growing crops for biomass energy production could further leverage that amount. Despite the emissions of carbon from destruction of forests and soils, it is worth remembering that the land system as a whole is currently acting as a net sink for carbon on a global scale, although it is unclear whether the land will ultimately turn into a source of GHG emissions as the planet warms in the 21st century.

Land Reflectivity

Human changes to the landscape have also altered how much of the incoming sunlight is reflected back toward the sky. A very white surface (such as a field of snow or ice) reflects a lot of light—while a very dark surface (such as a freshly plowed field of black dirt or the open ocean) absorbs a lot of light, ultimately heating Earth and its atmosphere. The scientific term for the reflectivity of a surface is albedo, and albedo values range from zero (absorbing all light) to one (reflecting all light).

People who swelter through hot, urban summers may be intuitively familiar with how albedo works. There are a number of reasons cities are often hotter than surrounding areas. One reason is that so much of the surface area of a city consists of black, tar roofs or blacktop roads and parking lots.

But reflectivity does not always work in obvious or intuitive ways, and the heat-absorbing properties of a surface may be influenced by a number of factors. We are used to thinking of trees as cooling us off with their shade—and they do keep us from being heated by direct sunlight—but those trees themselves absorb much of the energy they protect us from.

On a global average, human changes to the land surface are estimated to have a net cooling effect by raising albedo in more places than they lower it. This seems to be primarily the result of the clearing of trees. Deforested areas in the higher latitudes reflect a lot more light when covered with snow than do forested areas. There is considerable uncertainty about how strong this effect is, because quantitative information about human-caused land cover change is itself poor. But several estimates put the effect in the ballpark of 0.2 W/m^2—roughly on the order of certain aerosols or some of the minor greenhouse gases.

Greenhouse Gas Emissions

Increases in the atmospheric concentrations of greenhouse gases dwarf the other human influences on global climate. Unlike some other human impacts, the greenhouse gas increases all tend to push the Earth's climate in the same direction: warmer.

A more detailed discussion of the major greenhouse gases—their sources, chemistry, and sinks—is included in Chapter 3, Greenhouse Gases. The following will specifically address how they fit into the human species' overall influence on global climate. (Figure 17.)

CARBON DIOXIDE (CO_2)

Carbon dioxide (CO_2) has more total impact on climate than any other greenhouse gas emitted by human

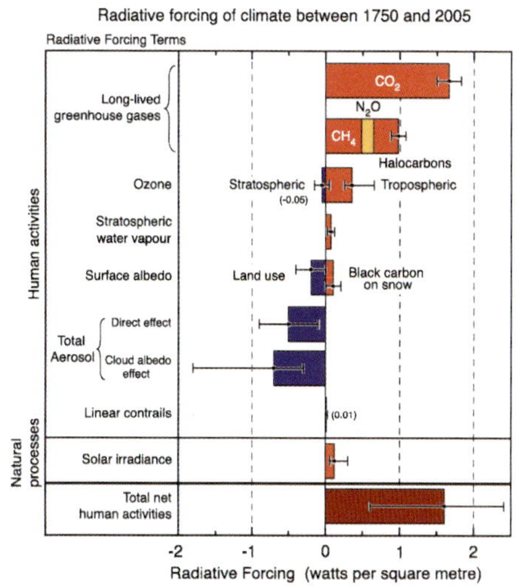

Figure 17. *This bar chart shows the global mean radiative forcings (and 90% confidence intervals) for GHGs and other effects. Positive values indicate warming, while negative values indicate cooling. The total net effect of all human activities is positive, about 1.6 W/m^2. Source: IPCC AR4 WGI, TS.5.*

activity. Since the Industrial Era started, humans have increased the concentration of CO_2 in the atmosphere by about one third, from about 280 ppm in 1750 to 379 ppm in 2005. This human-induced increase brings an additional radiative forcing of 1.66 W/m^2—compared to the natural solar forcing of 342 W/m^2.

The direct human emissions that have the most long-term effect come from combustion of fossil fuels like coal, oil, and gas, and from cement production. Such activities take carbon that has been locked beneath the Earth's surface for millions of years and release it to the atmosphere.

Anthropogenic emissions of carbon dioxide from fossil fuel combustion and cement production reached a peak of about 6.6 GtC/yr in 1997 (0.2 GtC/yr of that from cement production). CO_2 emissions continue on an upward trend, averaging about 7.2 GtC/yr from 2000–2005, compared to 6.4 GtC/yr over the 1990s and an average of 5.4 GtC/yr during the 1980s.

METHANE (CH_4) AND NITROUS OXIDE (N_2O)

Methane and nitrous oxide are next after carbon dioxide in the importance of their effect on climate change.

They are both well-mixed greenhouse gases with lifetimes measured in years to decades rather than centuries, produced by both natural sources and by industrial and agricultural activities. The increase above pre-industrial levels is estimated to add a forcing of 0.48 W/m^2 for methane and 0.16 W/m^2 for nitrous oxide—together little more than a third of the forcing from carbon dioxide.

TROPOSPHERIC OZONE

Various human industrial activities, including the burning of petroleum fuel in cars and trucks, contribute to the formation of smog, especially in cities during summertime. Ozone, the main component in smog, has harmful health effects when it occurs in the lower atmosphere (the **troposphere**)—that is, in the air we breathe. In addition, tropospheric ozone also has a significant greenhouse effect.

Quantifying this effect with confidence is very difficult, primarily because ozone is created and destroyed very quickly and has a very short atmospheric lifetime—a matter of weeks. Ozone concentrations in the troposphere vary widely over time and space, making it difficult to come up with meaningful global averages. Nonetheless, the best available model estimates suggest that anthropogenic increases in tropospheric ozone deliver an average forcing in the range of 0.25 to 0.65 W/m^2 and a best estimate of 0.35 W/m^2, putting it in a league with methane and nitrous oxide. By contrast, stratospheric ozone has a very slight cooling effect.

HALOGEN COMPOUNDS

The chlorofluorocarbons (CFCs) that most people are familiar with are only one of many families of compounds that contain the halogen elements (most importantly fluorine, chlorine, and bromine) and erode the Earth's protective layer of stratospheric ozone. (See Chapter 8, Stratospheric Ozone Depletion.) Many of these compounds also have a significant greenhouse effect.

Many of these compounds, especially the CFCs, have been banned under the Montreal Protocol, but their influence on the atmosphere will linger for decades or centuries after humans have stopped emitting them. Moreover, some of the "ozone-friendly" substitutes for them, such as of hydrochlorofluorocarbons (HCFCs), are potent greenhouse gases and remain in the atmosphere for hundreds or even thousands of years.

The halogen compounds differ from the other greenhouse gases in two important ways—their extremely long atmospheric lifetime and the much higher forcing effect they produce on a per-molecule basis. Thus their **global warming potential** (a measure based on their cumulative warming effect on a per-molecule basis over their lifetime) is much higher than for carbon dioxide, methane, and nitrous oxide (see Table 2).

The forcing effect produced by the two most harmful CFCs (already banned) at their current abundances in the atmosphere is 0.063 W/m^2 for CFC-11 and 0.17 W/m^2 for CFC-12. None of the others produces a forcing of more than 0.01 W/m^2, although there are scores of them.

AEROSOLS

Dust, smoke, and other forms of anthropogenic air pollution also have important greenhouse effects. **Aerosols**, as they are called, are microscopic solid particles or liquid droplets so light that they are suspended in the air, at least temporarily. (Figure 18).

The mechanisms by which various aerosols affect the Earth's radiative budget have only begun to be understood—and their importance appreciated—in the past decade or so. They are complex and conflicting (aerosols both warm and cool the Earth). Depending on the kind of particle, aerosols can either reflect or absorb radiation, either the shorter-wavelength incoming solar radiation (sunlight) or the longer-wavelength infrared radiation (heat given off by the Earth).

Estimates of their greenhouse effects are still fairly tentative. The net effect of aerosols is to shade the Earth—cooling it—scientists believe. But the situation is hardly that simple.

Anthropogenic aerosols that cool include sulfate aerosols, the main ingredient in acid rain. One of the

Figure 18. *This true-color image of a thick aerosol sulfates and organics plume, produced by the industrial northeastern section of the United States, is typical during summer months. Image taken May 4, 2001, by Sea-viewing Wide Field-of-view Sensor (Sea WiFS). Source: NASA.*

main pollutants given off by the combustion of coal and oil is sulfur dioxide (SO_2), a gas that is harmful to breathe. Once it has gone up a smokestack and into the atmosphere, sulfur dioxide combines with water to form sulfuric acid, and, through further chemical transformation, dry sulfate particles. These particles are suspended in the atmosphere as sulfate aerosols. They can fall back to the ground in the form of acid rain, which led to regulations in the 1990s requiring coal-fired power plants to scrub SO_2 from the gases released from smokestacks.

Sulfate aerosols are believed to have a shading (cooling) effect, reflecting incoming sunlight back to space before it warms the Earth or the atmosphere. Although sulfate aerosols travel continental distances, they are not globally mixed or uniform, and they eventually settle or wash out. Although certainty is low, sulfate aerosols are estimated to exert a negative forcing (i.e., cooling effect) of -0.4 W/m^2 (±0.2 W/m^2) on a global average.

Anthropogenic aerosols that warm include soot, one component of which is black carbon. Anyone who has seen the exhaust emissions from diesel trucks knows that burning fossil fuels can produce a dense, black smoke. That smoke includes lots of very small particles of elemental carbon. Without proper pollution control equipment, burning coal can also produce soot.

Black carbon is a very effective absorber of both sunlight and infrared, and as a result tends to warm the lower atmosphere (troposphere). Although its effects may be localized and transient, black carbon from fossil fuel combustion is estimated to cause a direct global average forcing on the order of 0.2 W/m^2 (±0.15 W/m^2).

Human activities lead to release of numerous other kinds of aerosols. Human burning of fossil fuels and biomass like wood produces many other compounds besides elemental black carbon. These carbon-based **organic** compounds take the form of both particles and gases. Fossil fuel organic carbon aerosols are estimated to produce a very slight cooling effect of -0.05 W/m^2 (±0.05 W/m^2).

Other significant aerosols produced by human activity include mineral dust and nitrates. The direct effect of nitrate aerosols, newly estimated in the AR4, is -0.1 W/m^2 (±0.1 W/m^2), and mineral dusts have been reevaluated to a lower value than in the TAR, -0.1 W/m^2 (±0.2 W/m^2). These estimates only include direct effects, and indirect effects considerably complicate estimating the net radiative forcing of these aerosols.

The AR4 estimates that, in total, all aerosols produce a direct radiative forcing effect of -0.5 W/m^2, with a range of -0.9 to -0.1 W/m^2. This slight cooling effect is better understood than in previous assessments, but still has a medium-to-low level of scientific understanding. The indirect effect of aerosols—primarily how they change the reflectivity of clouds—is even less well understood, but is currently estimated at -0.7 W/m^2, with a range of -1.8 to -0.3 W/m^2.

Global Change: The Growing Human Footprint

The ways in which humans have disturbed and changed the planet Earth during their short history upon it have been many and profound, as this chapter has suggested. Major human effects have included widespread deforestation; changes to grasslands, prairies, wetlands, peatlands, and the rich soils they contain; agriculture and the myriad changes it brings to soil, water, and air; and changes to the reflectivity (albedo) of vast swaths of the planet via these and other land-use changes.

We have looked at some of the direct ways humans have changed the atmosphere (and as a result, the climate) through emissions of greenhouse gases from a wide variety of activities. Carbon dioxide, methane, and nitrous oxide are the ones most often discussed—but beyond those gases are a host of other gases, such as tropospheric ozone and the halocarbons, which are generated by humans and also have significant climate forcing effect.

In the last decade, scientists studying climate have deepened their understanding of some other ways in which humans affect the planet and its climate. Aerosols are one of the most important, not only because of their direct heat-absorbing and light-reflecting properties, but because of their many complex and indirect effects such as promoting cloud formation.

These things are not science fiction. People are already changing climate on significant scales. Witness the Central Valley of California, an area where humans have built an extensive artificial irrigation network that has changed a desert into a breadbasket—and produced or intensified dense and dangerous ground fogs somewhat particular to the region. People living in large U.S. cities have gotten used to the idea that their regional climate is much hotter than surrounding areas (as well as more polluted), and we are now learning that cities have different rain and snow patterns as well.

As the Earth's human population continues to grow over the coming century, many of these effects will become more dramatic. As humans stretch the planet's resource base as far as it will go, we are also likely to find that we have become more vulnerable to some of the climate effects we have created.

CHAPTER 6: COMPLEXITY OF THE CLIMATE SYSTEM

Previous chapters have discussed what the complexity of the dynamic climate system really is: interlinking the Sun, atmosphere, oceans, ice cover, land, living things—and now humans. Some of the ways the different components of the climate system interact and how the system as a whole might react to the intensification of its natural greenhouse forcing have also been addressed.

Quantifying just *how much* human changes to atmospheric gases will influence climate is difficult for a number of reasons. Predicting the behavior of this dynamic, interlinked system of systems is more difficult because of the many **feedbacks** that influence its behavior when it is disturbed. The feedbacks can be both positive, amplifying the warming effect of human greenhouse forcing, or negative, counteracting the warming effect. Some of the feedbacks are sufficiently understood to assist in estimating climate change, while others are not. For a short explanation of the difference between feedbacks and forcings, see the box in Chapter 2, page 11.

It is comparatively easy for science to understand the simple and direct greenhouse effect—to understand the physics behind it, to reduce it to a mathematical equation, and to calculate what a specific change in greenhouse gases might cause in the way of temperature change—all else being equal. But of course all else is not equal.

The interlinkages, complex dynamics, and feedbacks are what really make the difference. Some of the factors involved in feedbacks that complicate things include water vapor, cloud dynamics, precipitation, oceans, land forms, and ice.

Water Vapor Feedback

As previously noted, water vapor (humidity) is actually the most powerful greenhouse gas in terms of its radiative effect—but again, it's not a radiative forcing. The amount of moisture in the air at a given place and time varies substantially.

The water-holding capacity of air depends on its temperature—warmer air holds more moisture. In terms of the most basic physics, this is a strong positive feedback. As climate gets warmer, the atmosphere will tend to hold more water vapor, warming climate further, and so on.

The AR4 states that changes in atmospheric water vapor dominate the feedback mechanisms that affect climate. Recent modeling and observational evidence strongly suggest a 50% amplification of mean global warming.

Cloud Dynamics and Feedback

Clouds both absorb and reflect radiation, and the formation of clouds depends on humidity, temperature, pressure (altitude), aerosols, and various other things. The physical properties of clouds change with changes in the climate; larger aerosol concentrations can make clouds more reflective, potentially altering the Earth's radiative budget—this is called the **cloud albedo effect**. Moreover, clouds are too small and evolve too quickly for current global climate models to represent from fundamental physical equations. Therefore, models simulate clouds through statistical submodels, called **parameterizations**, which predict the development and behavior of large cloud fields. Reducing uncertainties in cloud parameterizations is one of the most active areas of global change research.

The AR4 presented a range of values for the cloud albedo feedback: -0.3 to -1.8 W/m^2, with a best estimate of -0.7 W/m^2. All climate models indicate that the cloud albedo feedback is negative (cooling). Although progress has been made in modeling clouds (this estimate was not possible at the time of the TAR), the AR4 concludes, "Large uncertainties remain about how clouds might respond to global climate change." How different models handle cloud feedbacks, particularly for clouds, is the primary difference in predictions of the equilibrium climate sensitivity.

Since the publication of the AR4, evidence that the net cloud feedback is positive has continued to build. While a negative feedback cannot be eliminated as a possibility, recent studies suggest that the net feedback is likely positive (i.e., warming), at least for short-term changes in climate.

Precipitation Processes

The complexities of precipitation processes are similar in many ways to those of cloud feedbacks—and are also hard to quantify.

When water vapor condenses out of the atmosphere as rain or snow, heat energy (known as **latent heat)** is added to the air. The warmed air tends to rise, creating the familiar process of convection, which drives weather systems from thundershowers to hurricanes, redistributing heat energy. Individual rainstorms are too small to be represented in global models.

As temperature increases globally, so does evaporation. Globally, there must also be an increase in precipitation to balance the evaporation; but the precipitation need not occur in the same place as the evaporation, a significant complication. The release of latent heat during precipitation tends to intensify storms, further increasing precipitation—a feedback that suggests storms may become more intense in a warmer world.

Changes in precipitation can have various effects on other parts of the climate system. They can affect the amount of water in lakes and streams, the salinity of the ocean, soil moisture (which itself often feeds back to increase precipitation), or snow and ice cover. Moreover, precipitation can remove aerosols and soluble gases in the atmosphere, and those aerosols and gases themselves may have radiative effects.

Although deficiencies remain (particularly in the tropics), simulations of precipitation have improved since the TAR. Specifically, projected patterns of precipitation are better understood. The AR4 concludes that the amount of precipitation should increase at higher latitudes [>90% probability] but decrease in most subtropical (between 20° to 40° latitude) land areas [>66% probability]. Current research suggests that heavy rainfall events will increase in many regions, even in some places where the average rainfall is projected to decrease.

Ocean Feedbacks

Oceans cover about 70% of the Earth's surface and play profoundly important roles in the climate system. Some of those roles are briefly discussed here, but all must be well represented if climate models are to be realistic.

Because they have a much greater capacity to absorb heat than the atmosphere does, oceans act as a kind of "thermal flywheel," slowing the response of the climate system to changes in radiative forcing. The circulation of ocean currents plays a big role in transporting and redistributing heat energy globally. Also, much of the solar

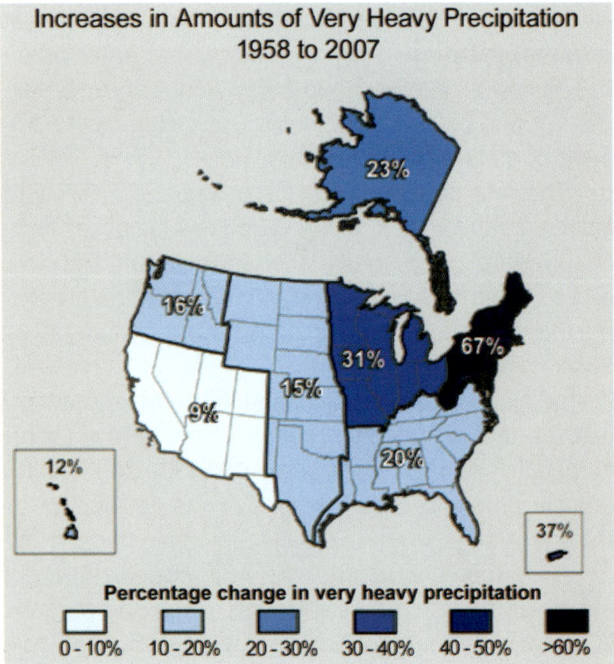

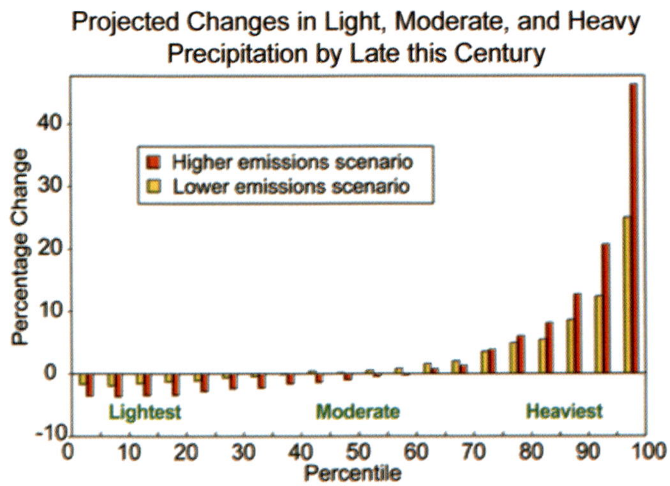

Figure 19. *The map shows the increase in the amount of precipitation in heavy downpours from 1958 to 2007. All regions of the United Stateshave experienced increases, but the change is particularly dramatic in the Northeast and Midwest. The bar chart shows model predictions for how precipitation may change under different emission scenarios. In both cases, scientists expect a decrease in the frequency of light precipitation and large increases in heavy precipitation. Light rain is gentler on crops and gardens, while heavy downpours can damage crops and cause erosion and flooding. Source: U.S. Global Change Research Program.*

EXTREME WEATHER EVENTS

Figure 20. *Heavy rains: might climate changes lead to more of them in the future?*

Extreme weather events include droughts, heat waves, heavy rainfall, floods, hurricanes, and tornadoes. These events can sometimes, but do not always, coincide with each other. Extreme weather events have the potential to cause a great deal of damage, which can carry heavy economic costs.

The effects of climate change on extreme weather events remain somewhat uncertain. There is some evidence that points to changes in the occurrences of these events, but there is no apparent pattern to these changes on a global scale, according to the Intergovernmental Panel on Climate Change. For example, in some regions, there is evidence of increased precipitation and increases in heavy and extreme precipitation events. Since the 1950s, research points to a 2–4% increase in the frequency of heavy precipitation events in the Northern Hemisphere. Furthermore, researchers believe that it is very likely that there will be more intense precipitation events over all land areas during the 21st century.

However, there have been no systematic changes observed in the frequency of tornadoes, thunderstorms, or hail events in areas that have been analyzed. Further research, on a global scale, is necessary to examine the possible effects of climate change on extreme weather events.

Following are descriptions of a number of extreme weather events.

Drought—a period of abnormally dry weather which persists long enough to produce a serious hydrologic imbalance (for example, crop damage, water supply shortage, etc.)

Heat Wave—a period of abnormally hot weather. The specific "parameters" of a heat wave vary from source to source. According to the Federal Emergency Management Agency (FEMA), a heat wave is a prolonged period when temperatures are at least 10 degrees higher than average temperatures for a region.

Heavy Precipitation—an abnormally large amount of rain, ice, or snow in a given period, often defined as more than two inches of water equivalent in a 24-hour period. Heavy precipitation can cause landslides; mudslides; soil erosion; damage to crops, natural resources, and property; and flooding.

Flood—high water flow or an overflow of rivers or streams from their natural or artificial banks, inundating adjacent low-lying areas. Floods can be caused by heavy rain, but they can also be caused by swift melting of ice or snow.

Hurricane—a rotating low-pressure system that includes thunderstorm activity and winds of at least 74 mph (64 knots). Hurricanes are categorized on a scale of 1 to 5 based on strength. Category 5 hurricanes, the strongest, have winds greater than 155 mph (135 knots), according to NOAA.

Tornado—a violently rotating column of air in contact with and extending between a convective cloud and the surface of the Earth.

energy absorbed by oceans goes to evaporate water—and this humidity feeds various atmospheric weather and climate systems.

In addition to their physical climate interactions, oceans can also play important roles in the geochemical cycling of carbon. Seawater both absorbs and releases carbon dioxide. As the concentration of CO_2 in the air increases in relation to the amount of CO_2 dissolved in water, the water absorbs more CO_2. But as water gets warmer, its ability to absorb and dissolve CO_2 decreases. This mechanism is the basis of changes in the direct forcing effect: as warmer climate warms the sea surface, less CO_2 will be absorbed, raising atmospheric CO_2 concentrations, and further warming the climate.

Figure 21. *North Pacific storm waves as seen from the M/V Noble Star, Winter 1989. Source: NOAA.*

Land Feedbacks and Complexities

A realistic and detailed description of the physical land surface is important to accurate climate modeling. When solar energy hits rock and soil directly, some of it is absorbed and transformed into heat energy, which is eventually transferred to the air. Just how this happens depends a lot on the type of surface—its albedo, roughness, heat capacity, etc. Moreover, the movement of winds is influenced by the contours of the land surface—mountains, valleys, vegetative cover, and more.

Added complexities arise in trying to account accurately for the role of vegetation in land-air interactions. Some of the heat energy absorbed and released by the land surface is transferred through plants. Furthermore, plants absorb solar energy through photosynthesis and release latent heat through evapotranspiration. Obviously, climatic changes in temperature and precipitation will change the plant physiology that controls these processes. One measure of success in modeling these biological processes is how well models can simulate the daily and seasonal effects of vegetation.

Figure 23. *Aerial image of B-15 iceberg that calved from the Ross Ice Shelf in Antarctica in late March 2000. Source: Josh Landis, National Science Foundation.*

Figure 22. *Smithsonian Environmental Research Center (SERC) experiments expose portions of salt marsh and forest ecosystems to elevated CO_2 concentrations in outdoor chambers. Source: Smithsonian Environmental Research Center.*

Finally, there is the question of the interaction between land-based plants and atmospheric concentrations of carbon dioxide. It has long been well known among commercial nursery growers that increased CO_2 concentration in a greenhouse stimulates the growth of certain plants—known as a **fertilization effect** (Figure 22). This increases the rate at which plants take CO_2 out of the atmosphere—a negative feedback (cooling). However, most experiments show that this effect is relatively short-lived and highly dependent on the availability of water and nutrients.

Ice Feedbacks

The portion of the part of the planet that consists largely of snow and ice is called the **cryosphere**. It includes the polar sea ice, continental ice sheets, permanent and seasonal sea ice, alpine glaciers, permafrost, clathrates, and seasonal snow cover. All of these interact with climate in important ways.

The most important cryospheric feedback is a result of the greater albedo of ice and snow, compared to other land and sea surfaces. Ice and snow absorb less solar energy and reflect more back to space. The colder the planet gets, the more ice and snow covers it, reducing the amount of surface heating and further cooling the planet.

The same feedback works in reverse—warmer climate reduces snow and ice cover, lowering albedo and further warming the climate.

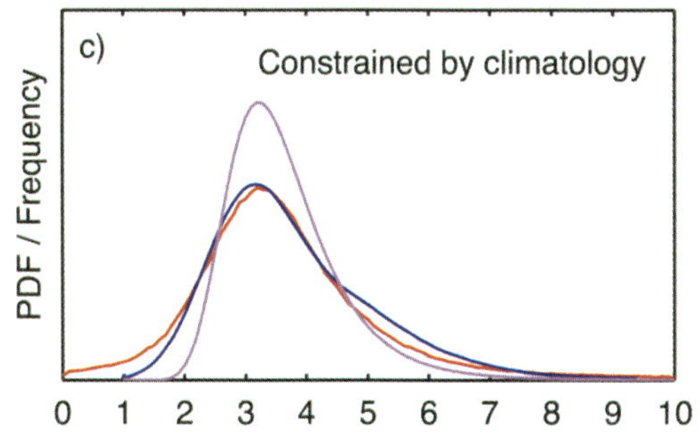

Figure 24. *This plot shows estimates of the probability distribution for the climate sensitivity from three different published climate papers. The peaks are all near 3° C (meaning this is the mostly likely value). Note that the curves are asymmetric—and that higher values are more probable than lower ones. Source: IPCC AR4 WGI, Box 10-2.*

Climate Sensitivity

The AR4 reports that progress has been made in understanding and modeling all of these feedbacks. So much so, that the AR4 reports significant improvements in understanding and quantifying the climate response to radiative forcing. Although several definitions exist, **climate sensitivity** generally refers to the average global warming that could be expected if atmospheric CO_2 levels reached and stabilized at *twice* their pre-industrial value (around 550 ppm) and the climate system reached equilibrium at this new state.

The TAR estimated the climate sensitivity at between 1.5°C and 4.5°C, but at that time it was not possible to make a best guess or even to estimate the probability it might be outside that range. The AR4 updated the likely [>66% probability] range for climate sensitivity to 2.0°C to 4.5°C, and provided the first ever best estimate of the

UPPER ATMOSPHERE VS. SURFACE TEMPERATURES

Different trends have been observed in the temperature at the Earth's surface, in the troposphere, and in the stratosphere. Temperatures observed at the Earth's surface over the past century show a warming trend. Specifically, surface temperatures have increased by about 0.6°C over the 20th century, and since 1979 (the start of satellite records), global average surface temperature has increased by about 0.15°C ± 0.05°C per decade.

However, in the troposphere, the layer of atmosphere closest to Earth, which extends to about 10 km (6 miles) above the surface of the Earth, temperatures also appear to be warming, but at a much slower rate.

In the low- to mid-troposphere, temperatures have increased at a rate of 0.05°C ± 0.10°C, much slower than the increase occurring at the surface. And in the upper troposphere, no warming has been observed since the 1960s. In the stratosphere, which extends from about 10 km above the Earth's surface to 50 km above the surface, there appears to be a cooling trend. This trend has been observed since the 1960s, with two sharp warming periods, each lasting for one to two years following the eruptions of the volcanoes El Chichon and Mt. Pinatubo in 1983 and 1992 respectively. The trends range from a decrease of 0.5 to 0.6°C per decade in the lower stratosphere to a decrease of 2.5°C in the upper stratosphere.

Those who disagree with the theories of climate change and global warming often cite the cooling of the stratosphere as evidence against global warming. They state that if the earth were actually warming, the stratosphere would be warming as well. However, evidence suggests that the cooling of the stratosphere is not incompatible with the theory of global warming. The downward trends in the stratosphere are consistent with models of the combined effects of ozone depletion and increased concentrations of greenhouse gases, mainly carbon dioxide and water vapor, according to the IPCC.

Although the eruptions of El Chichon and Mt. Pinatubo resulted in brief periods of stratospheric warming, the years following the initial warming show a return to cooling. (See Chapter 10, Stratospheric Ozone Depletion.)

climate sensitivity, 3°C. (Again, this means that *the best available science indicates that at 550 ppm (including all GHGs, not just CO_2), and assuming concentrations stay there, the global average temperature would be 3°C above the pre-industrial value.* We reached approximately 460 ppm in 2005.)

The AR4 was also able to put some probability estimates on the range for the climate sensitivity. It concluded that the climate sensitivity is very unlikely [<10% probability] to be below 1.5°C, and notes that values substantially higher than 4.5°C cannot be excluded. Importantly, the probability distribution is skewed toward higher values. (Think of a bell curve being shorter on the lower end and longer on the higher end, so that it's no longer symmetric.) Simply put, this means the eventual warming we can expect is more likely to be greater than 3°C than it is to be smaller than 3°C.

CHAPTER 7: MODELS AS WORKING REPRESENTATIONS OF REALITY

Modern society could experiment with climate by continuing to pump greenhouse gases into the atmosphere, but it might not want to accept the consequences in the end. Unfortunately, we do not have other Earths on which to conduct controlled experiments to determine what will happen. Computer models are valuable scientific tools for examining various climate hypotheses without harming the patient, so to speak.

Computer models can be used to investigate climatic complexity and variability. The computer simulations give scientists the opportunity to ask "What if?"—to examine possible consequences of actions before they cause significant change. Even with today's gargantuan supercomputers, however, no model can simulate the behavior of every wisp of cloud or every gallon of seawater.

It is worth remembering that a computer climate **model** is merely a simplified mathematical description of how scientists think the climate system works. A model is a set of working hypotheses or theories and is designed to encapsulate an understanding of the system in a quantitative way. Indeed, scientists often use the terms "theory" and "model" interchangeably. Computer models can only tell us the implications of what we think we already know about how natural climate processes work. But a model cannot "prove" a theory—a model must be verified with observations. Only by combining observations of a particular effect with model predictions based on our idea of what's happening, can we gain true understanding of the phenomenon.

There are many different types of climate models, ranging from the relatively simple (e.g., to examine a very specific aspect of climate) to the very complex, which attempt to factor in atmospheric and ocean influences around the Earth.

Simple Climate Model

In the simplest sense, a model is a rule or set of rules that attempts to describe how something in the real world works. For example, an equation in physics states that for a given confined mass of gas held at a constant temperature, its pressure is inversely proportional to its volume. Squeeze the gas into half the space, and its pressure doubles.

This equation is known as Boyle's law. In other words, if P is pressure, V is volume, and K is a constant, then $PV=K$. This equation provides a way of predicting how a confined gas will behave. As long as Boyle's law holds true, and as long as the temperature does not change, it is a good model. We could program a pocket calculator to tell us what P is when we punch in V, and vice versa.

One could devise a model attempting to describe what happens to solar energy arriving at the Earth. Such a model might be built on the assumption that incoming radiation must be balanced by outgoing radiation. This too could be expressed as an equation. We could add further refinements: How much incoming radiation is reflected by clouds and by the Earth's surface? And how much is absorbed and turned into heat? Each of these phenomena could be put into the equation as a numerical value—as a percentage, for example. The equation would work as a system: raise the amount of energy reflected by the Earth's surface, and you would have to reduce the amount absorbed accordingly. This is a simple numerical model.

Models such as this one could be called zero-dimensional. They treat all solar energy as if it were a single ray, or arrow, impinging on the Earth at various altitudes, and they describe only a single thing: energy. A more complicated model might try to take into account the fact that the Earth has a curved surface, that it is spinning, and that solar rays striking the equator might act differently from rays striking the poles. Such a model would add the spatial dimensions of latitude and longitude to the dimension of altitude. Such models have the advantage of allowing quick estimates of the climate response to a given emission scenario.

Three-Dimensional Models

Models used for weather and climate studies treat space three-dimensionally. They divide the Earth's atmosphere into multiple grid boxes and layers, then describe numerically what happens in each cell.

Most early climate study models were derived from a kind of model developed for weather forecasting, called general circulation models (GCMs). They attempt to describe flows of energy, air mass movement, and moisture over the entire globe. Much is understood about global circulation. Consequently, it is possible to construct a numerical model that describes it quite realistically. Climatologists know, for example, that the Sun heats the Earth's surface at the equator more than it does at the poles. And they know, too, that the redistribution of heat from the equator to the poles, together with Earth's spin, is what drives the major prevailing air currents, such as the jet stream and the trade winds.

All this can be expressed in the form of five major equations that describe the laws of conservation of momentum, heat, and mass. This system of equations is solved repeatedly for each grid square over successive time steps in a GCM. The models work well enough to be quite useful for weather prediction.

GCMs are really just a foundation on which researchers build more complex models for studying possible climate change. For example, some models try to account for soil moisture, sea ice, or other variables.

Coupled Atmosphere-Ocean General Circulation Models

Because the oceans serve as a giant reservoir, exchanging heat and moisture with the atmosphere, the coupling of atmospheric models to ocean circulation models is critical to making reasonable predictions. Three decades ago, climate models scarcely attempted to simulate realistically all the processes going on within the oceans and all the ways in which oceans and the atmosphere are linked in the climate system. Oceans were represented as monolithic "slabs"—without any allowance for their various layers, the mixing of water between those layers, dissolved salts and gases, or the circulation of their gigantic global currents.

GCMs are the most sophisticated and comprehensive climate models. They have improved dramatically since the TAR, and now over 20 different research groups from around the world run their own Atmosphere-Ocean GCMs (AOGCMs), which is valuable because it allows comparisons among model results. AOGCMs include dynamic processes representing the atmosphere, ocean, land, and sea ice. For even more complicated processes that are less well understood, like the formation of clouds and precipitation and the mixing of sea water due to ocean waves, parameterizations are still used—and uncertainties in these choices are the main reasons that the models produce different results.

Still, AOGCMs represent the frontier in climate modeling, and they are the primary tools not only for predicting future climate, but also for understanding and attributing past climate. Specific improvements include model formulation (better spatial resolution and improved parameterization), better simulation of present climate (such as surface temperatures and precipitation), and the ability to simulate El Niño, although other internal oscillations remain difficult to model.

Strengths and Weaknesses of Numerical Models

Meteorologists have developed such skill with computer models that they can use them to forecast weather quite successfully much of the time. These models are called numerical weather prediction (NWP) models, because they quantify heat, moisture, and wind momentum, using the same set of equations atmospheric GCMs use—but with finer resolution of space and time dimensions.

There are good reasons that numerical weather models work so well. They are based on the laws of physics. They are quantitative, and as a result they can sometimes be quite precise. The main distinction between NWP and climate models is that NWP models start with a known set of weather conditions. They generally look only a few days into the future, and weather forecasters can even assign probabilities to their predictions. Climate models, on the other hand, start with a "no weather" world and establish the climate from basic physical principles over a century or more, with no changes in radiative forcing until the system reaches equilibrium. These models then calculate 30-year averages of climate variables as they evolve under controlled changes in radiative forcing.

There are limits, however, to the usefulness of traditional numerical weather or atmospheric circulation models for predicting climate. For one thing, climate is influenced by many conditions and forces not much accounted for in weather models, such as ocean currents, land cover, ice, and myriad chemical and biological processes.

Climate models often have coarser resolution and are run forward much further into the future than traditional weather models. The further weather models look into the future, the more uncertain their predictions, because the atmosphere alone, and therefore the weather, cannot be predicted more than about two weeks in the future.

If long-term weather forecasts, on the order of a month ahead, have less than a 60% chance of accuracy, then how can a climate forecast 50 years into the future be reliable? In answering some questions, far-future modeling will be of no help at all. Will it rain or shine on a given April day in the year 2050? Models will not give a reliable answer. On the other hand, when the atmosphere is coupled to the more slowly changing components of the climate system, like the oceans and sea ice, the coupled models can give estimates on how the "average" weather will change. In answering other questions, however, the model may be quite reliable. How much will the average annual temperature at the equator change? The model may come much closer to answering that question.

As in journalism, the key lies in asking the right questions. Climate is about those major aspects of the atmospheric system that vary only slowly over time, even as weather varies daily. Climate, and climate modeling, is about averages and tendencies. On any timescale scientists can imagine, the Sun will always heat the poles less than it does the equator because it strikes the poles at oblique angles. Earth will always spin in the same direction. The continents and oceans will "always" (for purposes of useful current perspectives) be in the same position. These underlying forces do not vary randomly over time but will continue to influence climate 50 years into the future just as they do now.

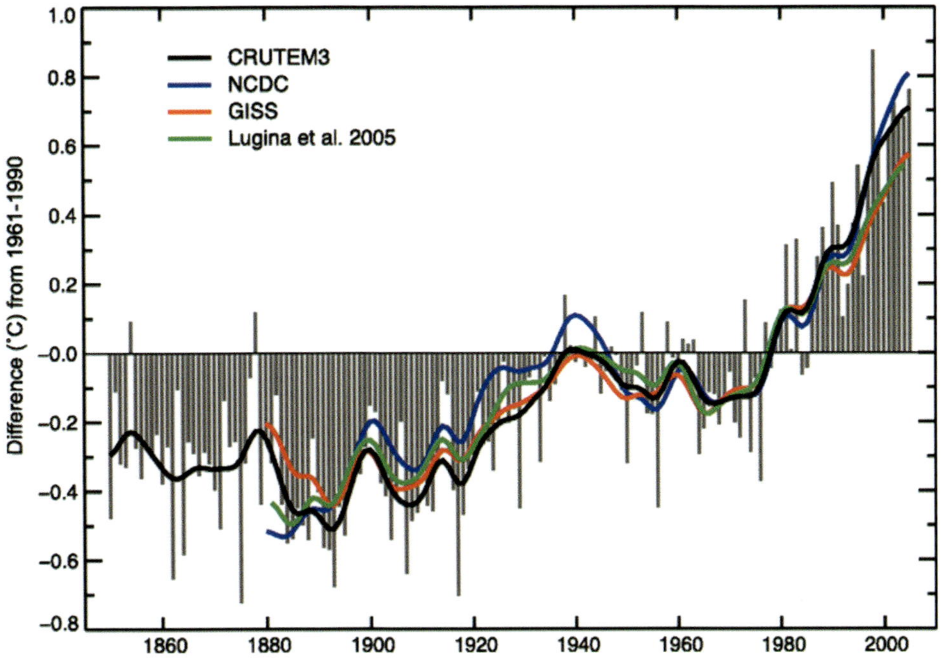

Figure 25. *This plot shows actual observations of surface temperature (relative to the period 1961–1990) from four different climate research groups. Observations clearly show an increasing trend over the past decade. Source: IPCC, AR4 WGI, Ch. 3, Figure 3.1.*

Validating Models

It is easy to succumb to the temptation of confusing models with the reality they are supposed to represent. But models are theories, not laws of nature. Models are tested with "reality checks" based on past climate.

Fortunately, the reality checks available to climatologists continue to improve. Systematic measurements were institutionalized by many governments during the 20th century. Additionally, since the late 1950s, the United States and other countries' satellites have offered unprecedented observation platforms for instruments studying weather and climate on the global scale. The current generation of instruments, such as NASA's Earth Observing System, is designed specifically to probe those aspects of the climate system where questions remain unanswered.

Still, satellite instruments remain only as good as their calibration. The vast amounts of data they can supply are valuable only if they are repeatedly validated and adjusted by comparing them with measurements from ground stations and instruments in the lower atmosphere and at sea. Those measurements will continue to be vital.

Historical and Paleoclimate Information

Journalists are veryl familiar with the strategy of asking a question whose answer they believe they already know. Similarly, one of the first ways to validate a climate model (or any theory) is to ask it a question to which you already know the answer. Knowing the historic surface temperatures and concentrations of greenhouse gases, scientists can see how well their models "predict" what they know to have actually already happened.

Past climate that predates the historical record, or what scientists call **paleoclimate**, offers a wealth of information that does more than just validate models. It supplies basic insights into climate processes and what can be expected from them. Unfortunately, the record of the Earth's climate history, as measured consistently by instruments in many places, is scarcely 140 years long (Figure 25). Consistent instrument measurements of CO_2 concentrations are just over five decades old.

Nonetheless, paleoclimatologists have collected a vast amount of indirect data about past climate through a number of other means. For example, written shipping records noting ice conditions in the canals of the Netherlands, or winemakers' records of grape harvests, can be used to infer long series of year-to-year variations. Tree-rings tell much about patterns of wet and dry (or cold and warm) years going back in history. Other sources yield data about even older climates: examination of sea-bottom sediments, fossil pollen grains, oxygen isotopes in fossil water, air bubbles in deep glacial ice samples, and so forth. Still older geological clues, carved in the topography of the rocks themselves, record the changing sea levels associated with the coming and going of ice ages over millions of years.

Inherent Predictability Limits

When statisticians talk about making a **prediction**, they mean something quite specific. A statement is not a prediction unless it has two parts: a statement about what might happen in the future, and a quantitative statement about the probability of that event's happening.

When meteorologists say there is a "65% chance of snow," they usually are making predictions, properly named, that are well grounded statistically. In fact, computer models of weather systems—combined with lots of real-life data about how accurate those models have been in the past—allow them to come up with a number for their chances of being right.

Climate models make predictions in a number of different but clever ways. For example, repeated runs of a single model will also yield differing results—which can also give a rough indication of the possible magnitude of the climate system's internal variability. One way to compensate for any potential blind spots of one particular model is to run many different models, with different conceptual underpinnings. The range of different model results gives some clue about the range of possible error. Furthermore, models can be run repeatedly using a variety of initial conditions and assumptions, and those model runs also yield meaningful information about the range of possible outcomes.

Undertaking efforts like these tends to increase modelers' confidence that a real-life climate outcome will fall within a certain range. It is fairly easy to pick a central value from within such a range, but because such central estimates cannot be assigned probabilities in a statistical sense, modelers tend to emphasize ranges rather than central estimates. It is important not to interpret these central estimates as predictions of the future.

Even if the climate system worked with machine-like predictability most of the time, it might sometimes behave unpredictably. It is possible that the climate system, instead of changing smoothly, might be unstable and change more abruptly, flipping into a new equilibrium state. In fact, there is some paleoclimatic evidence suggesting that past climate has produced such "surprises." Individual climate model runs, in fact, often show abrupt changes and offer clues on how regional climate might behave.

In terms of Earth science, scenarios of climate change inevitably involve some uncertainty about how most parts of the climate system behave. While this reality may disappoint those who would like to see only absolute certainty, models are consistently indicating a strong possibility of dramatic consequences, and thus can help inform our decisions on how to respond.

Predicting the Past: The 20th Century

The "final exam" in testing AOGCMs involves how well they simulate the observed global warming of the 20th century. After all, the 20th century is when human changes to the atmosphere became most pronounced, and when the quality of actual data on climate improved significantly.

Current AOGCMs are much more advanced than the previous generation—in part because of improved success in reproducing observed climate from the last century. The AR4 points to climate variables such as surface temperatures, temperature extremes, sea-ice extent, ocean heat content, and precipitation—all of which current models can simulate pretty well. Model representation matches past temperatures very well (see Figure 13), and scientists can use the models to deduce cause and effect (as discussed in Chapter 9 on attribution).

The AR4 sums up the progress of current models this way: "AOGCMs provide credible quantitative estimates of future climate change, particularly at continental and larger scales," although "confidence in these estimates is higher for some climate variables (e.g., temperature) than for others (e.g., precipitation)."

CHAPTER 8: PROJECTIONS OF FUTURE CLIMATE

Projecting the state of the climate system in the future is a huge challenge, not only because of all the physical complexities we have discussed in previous chapters, but also because it depends on predicting other things first. The most important of these is the growth of atmospheric greenhouse gas concentrations, which depend on future GHG emissions, which in turn depend on population growth, demographics, policy decisions, human behavior, etc. Decades of data give little reason to hope or believe that the growth trends in GHG concentrations, particularly CO_2, will level off by themselves, much less head downward.

The IPCC attempts to capture this uncertainty around future emissions trajectories by constructing a set of **scenarios** for use in making projections. The scenarios cover a range of possibilities and serve to set bounds on projections of future climate. These scenarios are described in the IPCC's Special Report on Emissions Scenarios; however, this report was published in 2000, and global GHG emissions have surpassed even the most pessimistic scenario in recent years, at least until the global economic crisis of 2008–09. The next IPCC report, the AR5, will use a new set of scenarios that is currently under development.

> **THE BOTTOM LINE ON PROJECTIONS OF FUTURE CLIMATE CHANGE**
>
> - The expected warming by 2100 ranges from 1.1°C to 6.4°C above the average for 1980–1999, and the projected warming is positive everywhere on the globe.
> - Sea level may rise from between 0.18 to 0.59 m (relative to the 1980–1999 average) by the end of the 21st century, but notably, this is an underestimate because it does not include contributions from the loss of ice from large, land-based ice sheets.
> - Heavy daily rainfall events are expected to increase in many regions—even in some places where the average amount of rainfall is expected to decrease.
> - Heat waves are expected to be more intense, longer-lasting, and more frequent.

The AR4 summarizes a number of changes and impacts that are to be expected in a warmer climate, and importantly, it places probabilities and confidence levels in these conclusions.

Committed Climate Change

It's important to note that even if atmospheric GHG concentrations were held constant today (that is, if emissions suddenly dropped to zero), the climate would continue to change for a few decades. This is because the oceans and ice sheets take a very long time to fully adjust to temperature changes. **Committed warming** refers to "the further change in global mean temperature after atmospheric composition, and hence radiative forcing, is held constant," and the same concept applies to other aspects of climate, notably sea level.

The AR4 finds that holding atmospheric composition constant at 2000 levels would imply a committed climate change of about 0.1°C over the next two decades, and about 0.6°C by 2100.

The AR4 offers the following large-scale projections for climate change by the end of the 21st century:

TEMPERATURE

Expected warming by 2100 ranges from 1.1°C to 6.4°C above the average for 1980–1999, depending on the emission scenario. Assessed ranges in uncertainty are actually larger than in the TAR because of a more complete range of models and inclusion of carbon cycle feedbacks. *The projected 21st-century warming is positive everywhere on the globe.* The greatest increases are expected on land and at Northern Hemisphere high latitudes, and the expected warming increases moving inland from the coasts.

SEA LEVEL RISE

Sea level is expected to rise from between 0.18 to 0.59 m (relative to the 1980–1999 average) by the end of the 21st century, again dependent on the emission scenario. Notably, these estimates do not include contributions from the changes in the rate of ice flow from large, land-based ice sheets. (See Chapter 8 on Sea-Level Rise for more details.) Also, sea-level rise is not expected to be uniform around the globe, mainly as a result of changes in ocean circulation, and could vary by as much as 25% of the global average. The AR4 highlights common features among model projections, including smaller than average sea-level rise in the Southern Ocean, larger than average in the Arctic, and a "band of pronounced sea level rise stretching across the southern Atlantic and Indian Oceans." Research published after the AR4 indicates greater than average sea level rise on both coasts of North America, contingent on the fate of Greenland and West Antarctica ice sheets.

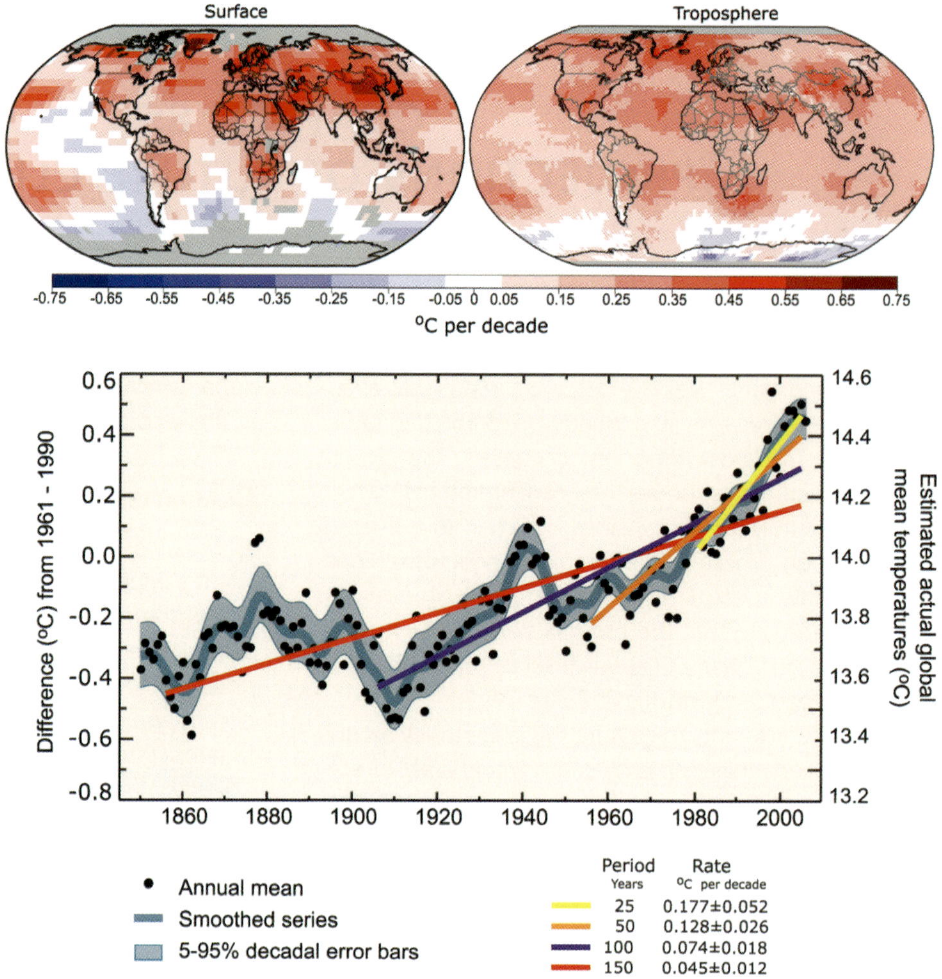

Figure 26. *The maps show the increase in temperature per decade at the surface (left) and in the lower atmosphere (right). The lower plot shows the global average temperature change (left axis) and actual average temperatures (right axis) since 1850. Source: IPCC AR4 WGI Technical Summary, Figure TS.6.*

SNOW AND ICE

Arctic sea ice is particularly sensitive to warming, and by 2100, large parts of the Arctic will lack year-round ice cover. Seasonal effects are strong; while projected changes in winter sea-ice extent are moderate, late-summer sea ice is projected to disappear entirely by 2100 under the higher emission scenarios. A few studies suggest this could happen considerably sooner. Snow cover is projected to decrease globally, and permafrost regions can expect widespread increases in thaw depths.

PRECIPITATION

The AR4 reports an improved understanding of projected precipitation patterns over the TAR. It finds, "Increases in the amount of precipitation are *very likely* [>90% probability] at high latitudes while decreases are *likely* [>66% probability] in most subtropical land regions." This is one area where models are being improved, but they do suggest that changes in precipitation will be harder to distinguish from natural causes than temperature increases.

EXTREMES

Models project an increase in heavy daily rainfall events in many regions—even in some places where the average amount of rainfall is expected to decrease. Heat waves are now better understood than in the TAR. Models suggest that heat waves are expected to be more intense, longer-lasting, and more frequent over the 21st century. Correspondingly, a decrease in the number of frost days is expected for the 21st century and in most regions, which could extend the growing season in some regions. Finally, although more research remains to be done, models suggest an increase in the number of intense hurricanes in a warmer future climate, but a decrease in the number of tropical cyclones globally.

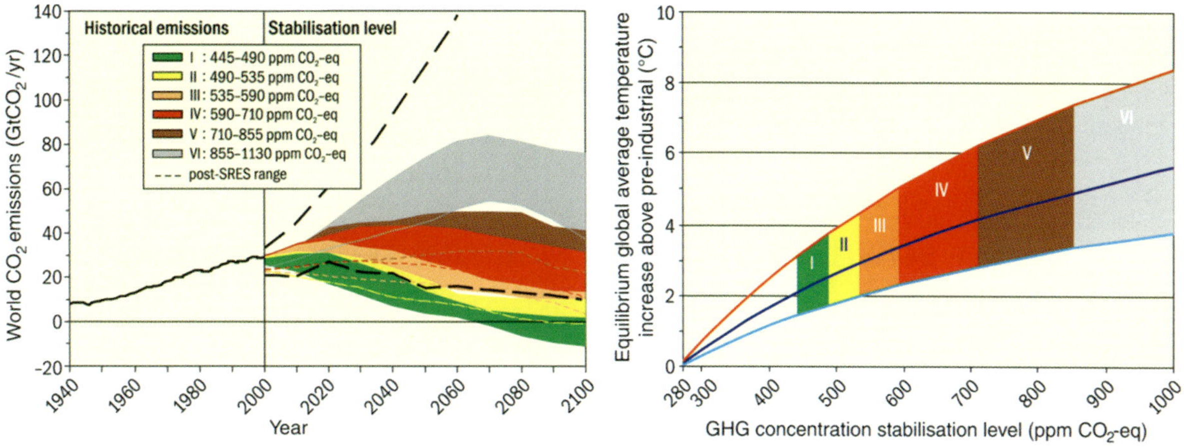

Figure 27. *These plots give the IPCC's projections for emissions growth through 2100 and the resulting impact on warming. Six scenarios (labeled I to VI, each in a different color) are shown in both plots. Green indicates the lowest emission scenario, where GHG concentrations level out between 445 and 490 ppm; the gray area indicates the worst case considered, where GHG concentrations land between 855 and 1130 ppm. The left plot shows the annual CO_2 emissions vs. time, and it gives an estimate of when worldwide emissions need to peak and start to decline in order to meet a given stabilization level. The right plot shows the corresponding range of temperature increase expected for each stabilization level. Even with the lowest stabilization level considered (green), the potential for significant warming (> 3 C°), and thus negative impacts, is real. Source: IPCC AR4 Synthesis Report, Figure 5.1.*

STABILIZATION

Figure 27 shows the IPCC's summary of what different stabilization levels would imply for global average temperature increases. The corresponding table provides ranges for temperature increases and sea-level rise under each stabilization level.

The left panel of the figure shows the projected CO_2 emissions through the end of the 21st century in order to reach the prescribed stabilization range in ppm. The colored shaded areas show emissions trajectories for a given stabilization target. The right panel converts these stabilization targets into a range of projected temperature increases. As is evident from the table, even reaching the lowest stabilization target assessed by the IPCC (445–490 ppm CO_2-eq, the green shaded areas) would imply significant warming (2.0°C to 2.4°C)—and to hit even this stabilization target

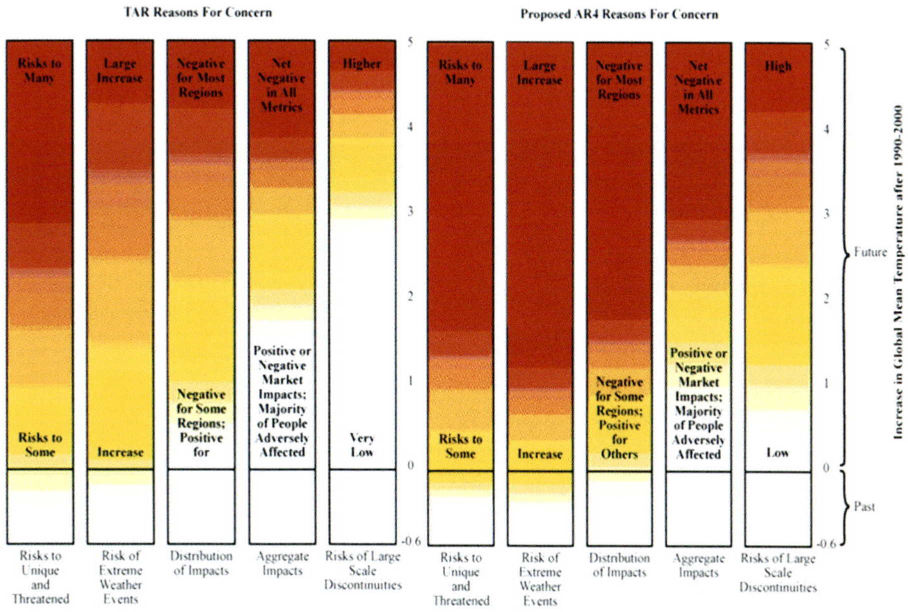

Figure 28. *These plots show qualitative measures of the severity of impacts due to climate change. The left side (five columns) appeared in the IPCC Third Assessment Report, and the right side (five corresponding columns) was prepared by some IPCC authors following the publication of the AR4 (Smith, 2009). Temperature is given on the vertical axis in degrees C, where 0 is present-day (1990–2000 average). Redder areas indicate more severe impacts. Source: Joel B. Smith et al.,* Assessing Dangerous Climate Change through an Update of the Intergovernmental Panel on Climate Change (IPCC) "Reasons for Concern," *106 Proc. Nat'l Acad. Sci. 4133, 4134 (2009).*

would require significant and immediate reductions in CO_2 emissions, as is clear from the left panel.

The IPCC Reasons for Concern diagram (Figure 28) gives a qualitative idea of what impacts can be expected under different levels of warming. This figure has been updated since the AR4 and, compared to the TAR, shows that impacts may be worse than anticipated.

DIFFERENCE BETWEEN DAYTIME AND NIGHTTIME TEMPERATURE TRENDS

Although daytime maximum temperatures have increased in recent years, nighttime minimum temperatures have increased at an even greater rate. Between 1950 and 1993, daytime maximum temperatures have increased by ~0.1°C per decade, and nighttime daily minimum temperatures have increased by about twice as much, ~0.2°C per decade.

Since nightly lows are increasing faster than daily highs, this results in a decreasing diurnal temperature range (the difference between the highest and lowest temperature in a 24-hour period). This diurnal range is also influenced by the type of land use with larger ranges in rural areas and smaller in urban areas.

As a result of the increase in nighttime minimum temperatures, the frost-free season has increased in certain regions, including those in the mid and high latitudes. The IPCC reports that these temperature trends are expected to continue, resulting in further reduced diurnal temperature ranges.

Furthermore, evidence indicates that there is also a difference between the daytime and nighttime trends in specific humidity, meaning the moisture (water vapor) content in the air. These trends are stronger—meaning there is a greater increase in specific humidity—at night than during the day.

Category	CO_2 concentration at stabilisation (2005 = 379 ppm)[b]	CO_2-equivalent concentration at stabilisation including GHGs and aerosols (2005=375 ppm)[b]	Peaking year for CO_2 emissions[a,c]	Change in global CO_2 emissions in 2050 (percent of 2000 emissions)[a,c]	Global average temperature increase above pre-industrial at equilibrium, using 'best estimate' climate sensitivity[d,e]	Global average sea level rise above pre-industrial at equilibrium from thermal expansion only[f]	Number of assessed scenarios
	ppm	ppm	year	percent	°C	metres	
I	350 – 400	445 – 490	2000 – 2015	-85 to -50	2.0 – 2.4	0.4 – 1.4	6
II	400 – 440	490 – 535	2000 – 2020	-60 to -30	2.4 – 2.8	0.5 – 1.7	18
III	440 – 485	535 – 590	2010 – 2030	-30 to +5	2.8 – 3.2	0.6 – 1.9	21
IV	485 – 570	590 – 710	2020 – 2060	+10 to +60	3.2 – 4.0	0.6 – 2.4	118
V	570 – 660	710 – 855	2050 – 2080	+25 to +85	4.0 – 4.9	0.8 – 2.9	9
VI	660 – 790	855 – 1130	2060 – 2090	+90 to +140	4.9 – 6.1	1.0 – 3.7	5

Notes:

a) The emission reductions to meet a particular stabilisation level reported in the mitigation studies assessed here might be underestimated due to missing carbon cycle feedbacks (see also Topic 2.3).

b) Atmospheric CO_2 concentrations were 379ppm in 2005. The best estimate of total CO_2-eq concentration in 2005 for all long-lived GHGs is about 455ppm, while the corresponding value including the net effect of all anthropogenic forcing agents is 375ppm CO_2-eq.

c) Ranges correspond to the 15th to 85th percentile of the post-TAR scenario distribution. CO_2 emissions are shown so multi-gas scenarios can be compared with CO_2-only scenarios (see Figure 2.1).

d) The best estimate of climate sensitivity is 3°C.

e) Note that global average temperature at equilibrium is different from expected global average temperature at the time of stabilisation of GHG concentrations due to the inertia of the climate system. For the majority of scenarios assessed, stabilisation of GHG concentrations occurs between 2100 and 2150 (see also Footnote 30).

f) Equilibrium sea level rise is for the contribution from ocean thermal expansion only and does not reach equilibrium for at least many centuries. These values have been estimated using relatively simple climate models (one low-resolution AOGCM and several EMICs based on the best estimate of 3°C climate sensitivity) and do not include contributions from melting ice sheets, glaciers and ice caps. Long-term thermal expansion is projected to result in 0.2 to 0.6m per degree Celsius of global average warming above pre-industrial. (AOGCM refers to Atmosphere-Ocean General Circulation Model and EMICs to Earth System Models of Intermediate Complexity.)

Table 3. Characteristics of post-TAR stablization scenarios and resulting long-term equilibrium global average temperature and the sea-level rise component from thermal expansion only.[a] Source: WGI 10.7; WG III Table TS.2, Table 3.10, Table SPM 5.

CHAPTER 9:
SEA LEVEL RISE

Over the past decade, predictions of future sea-level rise have changed based on evolving scientific understanding and newer research. According to the AR4, at a minimum the world may experience a globally averaged sea-level rise of 18 to 59 cm, covering the full range of IPCC emission scenarios. These estimates do not include an estimate for the change in ice flow from large ice sheets (Greenland and West Antarctica), since the IPCC concluded that the science was not well enough understood to make meaningful predictions.

Thus, the AR4 estimates include the contribution primarily from the thermal expansion of the oceans (water expands as it warms) as well as a small contribution from the melting of mountain glaciers worldwide. The causes of sea-level rise are discussed in more detail below. More recent research suggests that the AR4 likely underestimates the degree of 21st century sea level rise.

THE BOTTOM LINE ON SEA-LEVEL RISE

- Primary drivers of sea-level rise are thermal expansion of the oceans and melting of *land-based* ice; sea level is *not* affected by melting of sea ice or ice shelves.
- Sea level is expected to continue to rise through the 21st century and beyond.
- One outstanding question is the extent of the contribution from large land-based ice sheets.
- The measured rate of sea-level rise from tide gauges has been 1.8 ± 0.5 mm/yr for the period from 1961 to 2003, but more recent satellite observations indicate that rate has increased in recent years, showing a rate of 3.1 ± 0.7 mm/yr for 1993 to 2003; it remains unclear whether this increase is due to decadal variability or a true increase in the long-term trend.
- Current best estimates are between 2.3 and 3.3 feet of sea-level rise by 2100, if there is no action to mitigate greenhouse gas emissions; greater increases are expected on North American coasts.
- Estimates of future sea-level rise are higher than what is reported in the AR4 because research since that time has attempted to estimate the contribution from large, land-based ice sheets.

The IPCC's findings on sea-level changes described in this guide concern physical phenomena that can be observed, measured, and projected, and not the potential ecological or societal consequences of those changes in sea level, which is beyond the scope of this guide. For information on IPCC's conclusions on potential impacts to humans and on fragile coastal and marine ecosystems from saltwater intrusion, and extreme events such as floods and storms, see the separate report by IPCC Working Group II on Impacts, Adaptation and Vulnerability. (http://www.ipcc.ch/)

WHAT CAUSES SEA-LEVEL RISE?

The two major causes of sea-level rise are the **thermal expansion** of the oceans (that is, water expands as it warms) and the loss of *land-based* ice from increased melting of glaciers and ice sheets. These two processes alone cannot account for the entire observed rise, so other processes are also in play. These include thawing of frozen soils resulting in increased discharge from high-latitude rivers, and human-induced changes in land-based water storage (e.g., dams and extraction of groundwater), which is relatively poorly known. The AR4 concludes that the total balance of factors affecting global sea level is not yet sufficiently understood.

It's important to emphasize that ice already floating in water (such as the Arctic ice or various ice shelves) has no influence on sea-level rise as it melts. That's because the oceans have already been displaced to account for the ice volume (which is Archimedes' Principle)—like ice cubes in a drink. It's the *land-based* ice that is key for sea-level rise; as more icebergs and meltwater are dumped from the land into the ocean, sea levels rise. Mountain glaciers are particularly susceptible to increased temperatures, as evidenced by the widespread retreat of glaciers in all regions worldwide. Change in

Arctic sea-ice area is a key feedback in the climate system, and its recent decline is alarming, but the loss of sea ice has no impact on sea level.

Like most substances, ocean water expands when it warms. Both model results and observations point to thermal expansion as a major contributor to past and potential future sea-level rise. Because deep ocean temperatures change slowly, thermal expansion will continue for centuries, even if greenhouse gases and their effect on climate were to stabilize in the short term. This time lag also adds uncertainty to short-term projections on the timing of anthropogenic sea-level rise.

Climate change has a major influence on sea-level change, but other processes also have an impact. These include such things as storage of water on land (by dams, extraction of groundwater, modifications to surface characteristics affecting runoff, etc.); buildup of sediments at delta regions; vertical land movements resulting from geological processes and human activities; and changes in atmospheric and ocean dynamics. Terrestrial water storage, the IPCC suggests, may mask a significant fraction of the ocean volume increases resulting from thermal expansion and glacial melt. One kind of adjustment that is still occurring is referred to as "glacial rebound," resulting from weight shifts associated with the melting of large ice sheets. Tectonic adjustments can be rapid (like earthquakes), or gradual, like those associated with sediment transport.

The AR4 states that sea level shows "considerable regional variability" based on geographical variation of thermal expansion, salinity, winds, and ocean circulation. Compared to the global average, this range is significant and is critical when projecting how sea-level rise will affect a given location along a coast. More recent studies published after the AR4 indicate that the Pacific Northwest and the northeastern coasts of North America will experience greater sea-level rise than the global average.

How is Sea Level Measured?

Global average sea level refers to the level of the sea surface relative to that of the land, as measured by tidal gauges. These measurements contain information both on displacement of the land, and on changes in ocean volume. Present-day sea level is measured through two different techniques: tide gauges and satellite altimetry. Tide gauges provide data on sea level relative to the land on which they rest; data from tide gauge data therefore must be corrected for local vertical land motion. This motion can be caused from subsidence of the land or from glacial rebound, which is the reflex motion of landmass that was once covered with large (and heavy) ice sheets. Glacial rebound can be calculated from models, and although contemporary subsidence can be measured, it often is not recorded, so tide gauge data must be selected carefully (excluding areas with significant tectonic activity, for example). There are also concerns over geographical bias within tide gauge data, since most of the stations are located in the Northern Hemisphere and along coastlines (so that mid-ocean data are largely unavailable except on sparsely distributed islands).

Satellite measurements of sea level are more accurate but began only in 1993, whereas tide gauge data date back to 1870. Measurements from orbit are largely unaffected by land movements. The satellite observations have improved the estimates of sea-level rise and revealed complex geographical patterns of sea-level change in the open oceans.

Twentieth-Century Sea-Level Rise

Combining tide gauge data and satellite altimetry, the AR4 assesses the rate of global average sea level rise for 1961 to 2003 as 1.8 ± 0.5 mm/yr, and for the 20th century overall as 1.7 ± 0.5 mm/yr. See Figure 29.

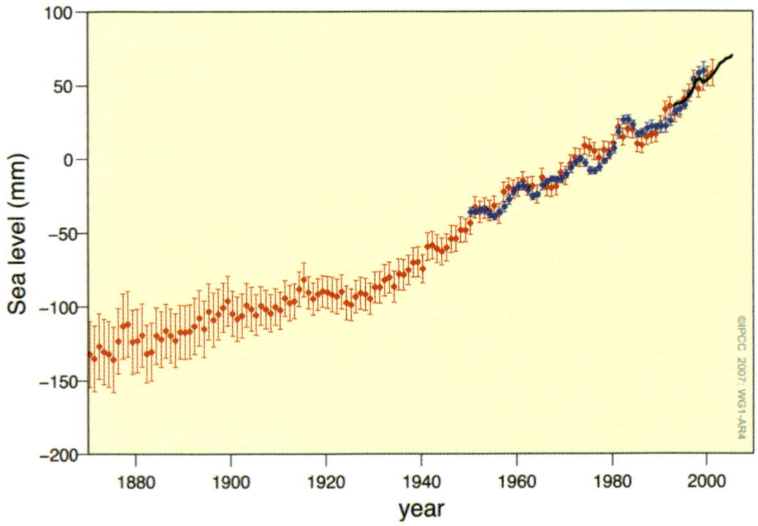

Figure 29. *Observations of annual global mean sea-level rise (relative to the average of the period 1961–1990). Red points are reconstructed sea-level fields, blue points are tide gauge measurements since 1950, and black represents data from satellite altimetry. Source: IPCC AR4 WGI, Technical Summary, Figure TS.18.*

Predictions for 21st-Century Sea-Level Rise

The AR4 indicates a projected range for the global average sea level rise by 2100 as 0.18 to 0.59 m. However, this result, according to the AR4, "does not assess the likelihood, nor provide a best estimate or an upper bound for sea level rise." The primary reason is that the estimate does not include future changes in ice from from large polar ice sheets (West Antarctica and Greenland). This reflected the authors' expert judgment that no conclusions could be drawn based on then-current observations and modeling; it did *not* indicate that these contributions could be neglected. As a result, experts generally believe that the AR4 likely underestimated the projected range of sea-level rise by the end of this century.

The magnitude of expected melting from the Greenland Ice Sheet and the West Antarctic Ice Sheet are a key uncertainty in projections of future sea-level rise, and this question is directly relevant to decisionmakers. Researchers continue to tackle this problem, and papers published after the AR4 attempt to estimate the contribution from large ice sheets. Two recent studies attempted to capture the ice contribution to future sea-level rise more completely than the IPCC approach, each using different methods. Projections in these two studies ranged from 0.5 to 2.0 meters (1.64 to 6.56 feet) for the end of the 21st century. Two other studies found that sea level does not rise uniformly around the world, and that the Pacific and Atlantic coasts of the United States will experience significantly more sea-level rise than the global average.

These conclusions were reaffirmed by the U.S. Global Change Research Program (GCRP), which released a report in mid-2009. The report extended the work of the AR4 to include more recent research, and it also was the first assessment to focus primarily on impacts and threats to the United States. It concluded that estimates of global sea-level rise have increased over the AR4 to between 2.3 and 3.3 feet, depending on the greenhouse gas emissions path. It also concluded that some areas of the United States (particularly the Atlantic and Gulf Coasts) will experience higher sea-level rise than the global average.

Finally, current thinking suggests that the chances of abrupt and large-scale sea-level rise may be greater than previously thought. One recent study found that during the last warm interglacial period about 120,000 years ago, sea level rose about 1.5 meters (4.9 feet) per century on average. During that time, a period of ice-sheet instability caused sea level along Mexico's Yucatán Peninsula to jump 2 to 3 meters in less than a century—and the Earth's average global temperature then was only 1°C (0.6°F) warmer than today.

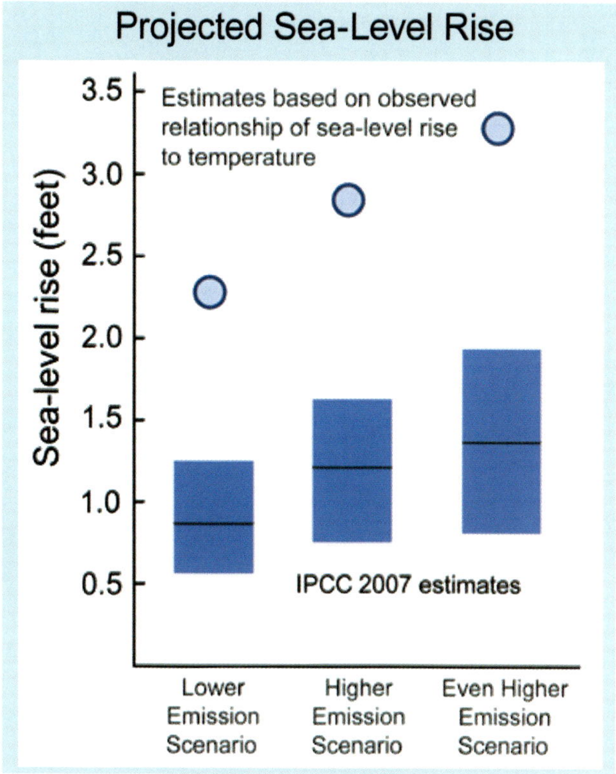

Figure 30. *Based on the state of the science at the time, in 2007 the Intergovernmental Panel on Climate Change (IPCC) projected a rise of the world's oceans from 8 inches to 2 feet by the end of this century (range shown as bars). However, they could not quantify the contributions to sea-level rise due to changes in ice-sheet dynamics. More recent research has attempted to quantify this contribution by estimating future sea level based on its observed relationship to temperature. For example, the projections indicated by the light blue circles in the figure above estimate global average sea-level rise of almost 3.5 feet by the end of this century under a high-emissions scenario. In areas where the land is sinking, such as the Atlantic and Gulf Coasts of the United States, sea-level rise will be higher than the global average. Source: U.S. Global Change Research Program, Climate Change Impacts in the United States (June 2009).*

Longer Term Outlook

Global sea levels will continue to rise after 2100 due to the time-lag effect in ocean heating. How much it ultimately rises and when it finally stabilizes is difficult to assess, but in large part depends on decisions made now to curb greenhouse gas emissions.

CHAPTER 10: STRATOSPHERIC OZONE DEPLETION

Introduction

While stratospheric ozone depletion and climate change are distinct environmental issues, they are also interrelated, because ozone-depleting substances (ODSs) are greenhouse gases and because the hydrofluorocarbons (HFCs) that replaced a portion of ODSs are also GHGs. Therefore, this chapter will present a brief description of stratospheric ozone depletion and its relationships to and with climate change.

The term "ozone hole" by now is widely recognized, but is it still news? Has the ozone depletion problem been solved? Due to the success of the Montreal Protocol on Substances that Deplete the Ozone Layer, the rate of ozone deterioration has declined in recent years, and the overall concentration of chlorofluorocarbons in the atmosphere has leveled off and is decreasing. But long-term recovery of the ozone layer will take many decades because many of the ODSs persist in the atmosphere for more than 100 years.

Although global and regional ozone depletion is no longer increasing, seasonal and annual fluctuations occur in the amount of observed ozone. Projections of future ozone amounts show that it will be a number of years before we have scientific confirmation of a sustained recovery, decades more before the ozone layer recovers to the condition when the ozone hole emerged, and a century or more before the ozone layer fully recovers from ODSs, and these chemicals are gone from the stratosphere.

The Montreal Protocol is recognized as the most successful multilateral environmental agreement, for the following reasons:

- A near complete phaseout of almost 100 ozone-depleting substances (chlorofluorocarbons, carbon tetrachloride, halons, etc.) was achieved in the past 25 years, resulting in clear scientific evidence that depletion of the earth's ozone layer has been arrested and appears to be on a long-term path to recovery.
- Elimination of these ozone-depleting substances, which are also potent greenhouse gases, has produced substantial climate cobenefits, buying time for governments to take effective action on reducing carbon dioxide emissions;
- The Montreal Protocol is unprecedented in participation and compliance; every country in the world participates as a member, with near-perfect compliance over two decades.
- Approximately $3 billion of investment so far has fueled global market transformation with little adverse economic impact, and no unwanted changes in lifestyle; and
- Lean and effective institutions and supporting networks were created that are respected by all governments and environmental and industry stakeholders; these networks can support climate change and other treaties.

The Protocol has obligations for emission reductions for both developed and developing countries, provides financing for the incremental costs of developing countries, contains compliance assistance backed up with necessary trade controls, and relies upon United Nations institutions that execute their responsibilities in a pragmatic manner in close cooperation with national authorities.

> ***The example of the Montreal Protocol sends a powerful message that action on major global challenges is not only possible, but that the financial and human benefits invariably outweigh the costs.***
>
> United Nations Secretary-General
> Ban Ki-Moon

The Ozone Layer—Protecting Life on Earth

Ozone is a pungent, highly reactive gas. It is a molecule made up of three oxygen atoms (O_3), and thus is a close chemical cousin to the more stable and abundant oxygen (O_2) needed for human respiration. Ozone is formed when a two-atom oxygen molecule is separated as a result of absorbing ultraviolet radiation from the Sun. The individual atoms of oxygen combine with

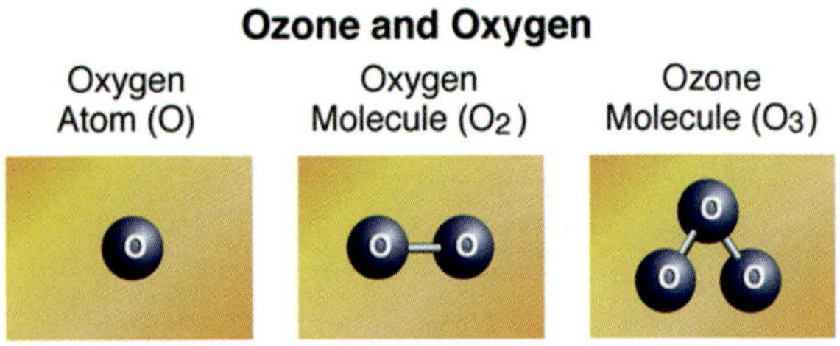

Figure 31. *Source: Scientific Assessment of Ozone Depletion: 2002, World Meterological Organization, 2003.*

individual molecules of oxygen ($O + O_2 = O_3$) to form two ozone molecules. (Figure 31.)

A critical point in understanding air pollution and stratospheric ozone issues is the distinction between ozone in the **stratosphere** (about 6 to 30 miles above the Earth) and ozone at ground level (in the **troposphere**). (Figure 32.) Ozone in the stratosphere, "good ozone," protects living things from harmful ultraviolet (UV) radiation from the Sun. Without that protection, the full radiation of the Sun would increase skin cancer and cataracts, suppress the human immune system, and destroy crops and ecosystems. (See box on UV Radiation.)

Exposure to ozone at ground level can be harmful to people, plants, animals, and buildings and other structures. Ground-level ozone is recognizable as the primary component in the smog that plagues many major urban areas, as well as rural areas with industrial facilities and power plants. (There are other common smog constituents, such as particulates, NO_2, SO_2, CO, and acid aerosols.)

Even though ozone molecules in the stratosphere play a critical role in screening the sun's ultraviolet radiation, they are exceptionally rare in the Earth's atmosphere. Even in the part of the stratosphere where ozone is most concentrated, there are typically only 1 to 10 parts of ozone per million parts of air, compared with about 210,000 parts of oxygen per million parts of air. If all the atmospheric ozone were moved to Earth's surface, it would occupy a layer about the thickness of two stacked pennies (about three millimeters). This low ozone ratio both underscores and belies the critical role ozone plays in protecting the global environment.

The amount of ozone in the stratosphere varies at different latitudes and altitudes, at different times of the year, and from year to year. These natural variations, which occur on daily to seasonal timescales, are caused by regular air motions and change in the balance of ozone production and destruction in chemical reactions caused by sunlight.

WHAT IS A DOBSON UNIT?

A Dobson unit measures the total amount of ozone in a column of air (total column ozone) from ground level to the top of the atmosphere. The number of Dobson units corresponds directly with the "thickness" of the ozone layer. If 100 Dobson units of ozone in a column of air were brought to Earth's surface, it would form a layer one millimeter thick. Total ozone values vary widely over the globe, typically in the range of 200 to 500 DU, with the lowest values in tropical regions and highest values in polar regions.

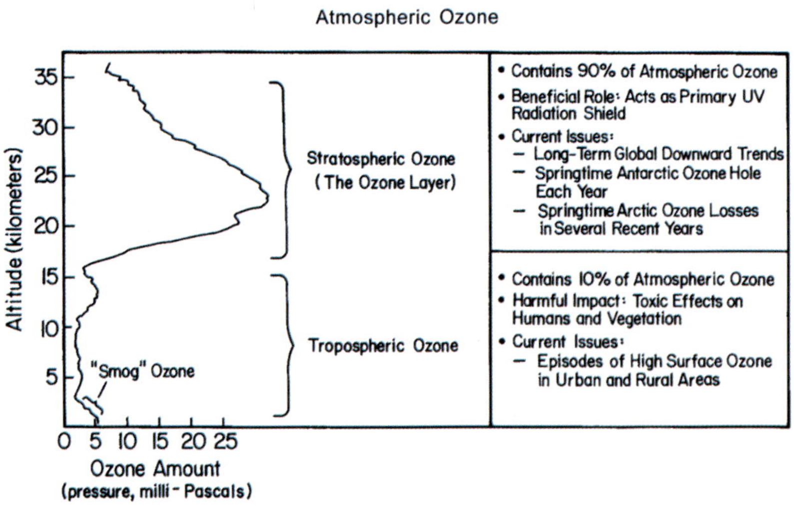

Figure 32. *Ozone in the Earth's Atmosphere. Source: World Meterological Organization, Scientific Assessment of Ozone Depletion: 1998, Global Ozone Research and Monitoring Project—Report No. 44, Geneva, 1999.*

Some of the ways that ozone molecules are destroyed occur naturally. For example, ozone molecules are destroyed by ultraviolet light and by naturally occurring compounds containing nitrogen, hydrogen, and chlorine. The nitrogen comes from soils and the oceans, the hydrogen mostly from atmospheric water vapor and methane, and the chlorine from the terrestrial and aquatic systems.

Until the 1980s, these naturally occurring forces that created and destroyed ozone were in balance in the stratosphere, with the average, long-term amount of ozone in the stratosphere remaining fairly constant. The scientific and public policy issue surrounding stratospheric ozone

ULTRAVIOLET RADIATION

While visible light has wavelengths between 400 and 700 nanometers (the unit equal to one billionth of a meter), UV radiation has wavelengths between 200 and 400 nanometers. UV radiation is of three types: UV-A (315 to 400 nanometers), UV-B (280 to 315 nanometers), and UV-C (100 to 280 nanometers). The shorter the wavelength, the greater the potential for harm. UV-A is beneficial in the production of vitamin D in humans but can also cause sunburn. UV-B can cause sunburn and is more biologically damaging (see section on Health Impacts, page 59), but fortunately, when the ozone layer is undepleted, most UV-B is absorbed by ozone in the atmosphere before reaching the Earth. UV-C is potentially the most damaging, but all of it is absorbed in the atmosphere, and it doesn't reach the Earth.

arose as a result of scientists' discovery that human activities—the releases of chlorofluorocarbons (CFCs) and halons (fluorocarbons containing bromine) produced by people—had upset the natural balance, and that ozone was being destroyed faster than it is created.

The Chemistry of CFCs and Stratospheric Ozone

The adage that "If something sounds too good to be true, it usually is" applies to chloroflurocarbons. After their discovery and commercialization in the 1930s and later, CFCs became an exceptionally useful and practical family of chemical compounds.

They are relatively inexpensive, highly effective in many applications, chemically stable at ground level, generally not flammable, and low in toxicity. These very technical properties and characteristics led to their widespread use for a wide variety of industrial purposes, including as solvents in electronics and aerospace manufacturing; refrigerants in air conditioning, refrigeration, and industrial heat transfer; and propellants in convenience or cosmetic aerosol products such as hairspray, deodorant, and pesticides.

But over time the "too good to be true" became more than apparent. All the CFCs contain the halogen species chlorine and fluorine. When released in the atmosphere, CFCs are transported by natural air motions to the stratosphere, where they absorb ultraviolet radiation, leading to the release of their chlorine atom. Chlorine atoms act as a catalyst, repeatedly reacting to break apart ozone molecules to re-form as an ordinary oxygen molecule and a chlorine monoxide molecule. The chlorine monoxide molecule then can combine with an oxygen atom, forming an oxygen molecule and freeing the chlorine to begin the process all over again. (Figure 33.) The process repeats over and over, destroying more ozone molecules.

Through this repetitive cycle, a single chlorine atom can destroy thousands of ozone molecules before it is chemically neutralized. Thus, CFCs in the stratosphere are powerful agents for destroying stratospheric ozone.

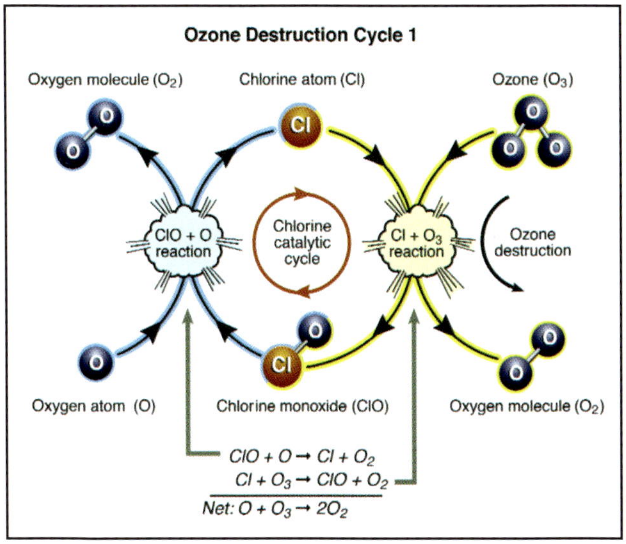

Figure 33. *Chlorine reacts with ozone (O_3) and oxygen atoms (O) to breakdown ozone and produce oxygen (O_2). Source: Scientific Assessment of Ozone Depletion: 2002, World Meterological Organization, 2003.*

In addition to chlorine, the other principal ozone-depleting chemical is bromine, which is released by halons used in fire-fighting and by methyl bromide used as a pesticide fumigant. Also long-lived in the stratosphere, halons are even more potent than CFCs in their **ozone-depleting potential** (ODP). Because of its shorter atmospheric lifetime, methyl bromide is less potent than CFCs, but is becoming a significant portion of ozone depletion as CFCs, halons, and other ODSs are phased out. Overall there is much less bromine than chlorine in the stratosphere, but bromine is more reactive and accounts for a disproportionate amount of ozone depletion. Methyl bromide emissions for quarantine and preshipment uses are currently not controlled by the Montreal Protocol and are increasing rapidly.

Under the Protocol (see discussion below), each covered chemical is assigned an ODP, a measure of its relative ability to destroy ozone molecules in the stratosphere. CFC-11 and CFC-12, with assigned ODPs of 1, are used as the reference compounds for establishing the ODP of other compounds. A chemical's ODP is determined by the number of chlorine or bromine atoms in the molecule and its atmospheric lifetime.

Sources of CFC Releases

Beginning in the 1930s, the use of CFCs expanded significantly, with production increasing at an average rate of 10% annually over the succeeding three decades. Among the products using CFCs and halons prior to the imposition of use restrictions and bans under the Montreal Protocol were:

- Rigid foams—used for insulation and packaging
- Flexible foams—used in furniture, bedding, and car seats
- Food and process refrigeration—residential, commercial, and industrial
- Air conditioners—motor vehicle, commercial, and residential
- Solvents—for cleaning electronic circuit boards, aerospace components and systems, and a wide range of other parts and assemblies
- Hospital sterilants, blood substitutes, lubricants for surgical needles, and syringes
- Fire extinguishers and explosion inerting (using halons)
- Propellants for medicines for asthma and chronic obstructive pulmonary disease, and
- Propellants in cosmetic and convenience aerosol products

As discussed below, aerosol products had been the single largest global use of ODSs prior to the mid-1970s when the United States, Canada, and Sweden banned the use of CFCs as propellants in nonessential aerosol sprays.

Landmark Discoveries

In 1974, two University of California–Irvine chemists, Mario Molina and F. Sherwood Rowland, published seminal research on stratospheric ozone depletion. Using laboratory experiments, they showed that CFCs, known to be highly stable gases, would rise to and eventually decompose in the stratosphere, freeing chlorine atoms to destroy ozone. Molina and Rowland's theory was initially challenged but eventually was confirmed by other scientists. In 1995 Mario Molina and Sherwood Rowland shared the Nobel Prize for Chemistry with Paul Crutzen for discoveries concerning ozone depletion. The CFC industry eventually came around to accepting the implications for continued release of ODSs and ultimately supported their phaseout.

The ozone-depletion issue first became widely known in the mid-1970s and remerged as a major issue in the mid-1980s with the discovery of the "Antarctic ozone hole." The term derives from satellite images of total ozone, which reveal a circular, continental-scale depletion of ozone over Antarctica in later winter and early spring. Thus, the ozone hole is a seasonal "thinning" of the ozone layer in the area over Antarctica. (Figure 34.)

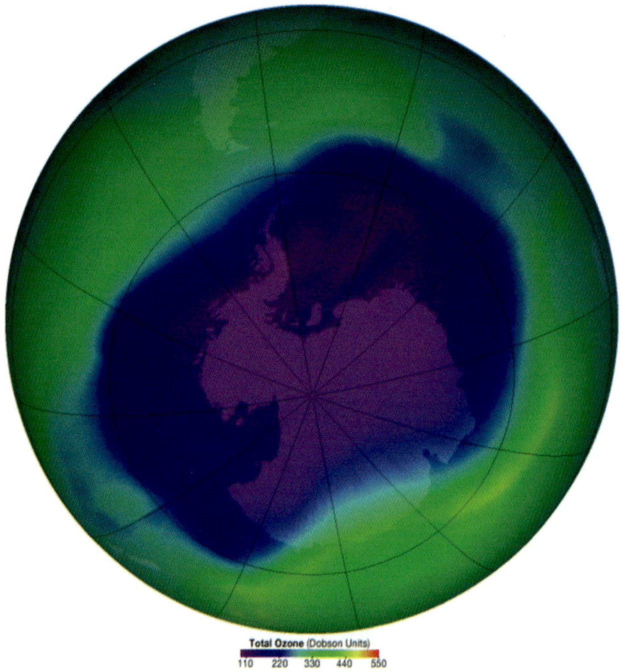

Figure 34. *Largest Ozone Hole Ever Observed, September 2006. Source: NASA, http://ozonewatch.gsfc.nasa.gov/ozone_maps.*

The Montreal Protocol

The Molina/Rowland warning regarding CFCs and the threat of stratospheric ozone depletion led to a rapid consumer response and subsequent public-policy response. Actions in the United States included boycotts of CFC hairspray and deodorant, promotion of CFC-free alternatives by competing companies, and state bans on widely used CFC-based aerosol products. Those state and consumer actions had nearly eliminated CFC aerosol products when, in May 1977, three federal agencies—the Consumer Product Safety Commission, the Environmental Protection Agency, and the Food and Drug Administration—announced a timetable for phasing out nonessential uses of CFCs in aerosol spray products. Canada, Denmark, Sweden and a few other countries followed, but other European authorities and citizens never took much interest in either boycotts or prohibitions.

By the mid-80s, world attention clearly had focused on the issue, and an international agreement, the 1985 Vienna Convention, was signed under the auspices of the United Nations Environment Programme (UNEP). The Vienna Convention established mechanisms for international cooperation in research and monitoring, and set a framework for international negotiations on emission reductions, but countries at this time could not reach agreement on measures to reduce emissions.

Two years later, in September 1987, sixteen countries including the United States signed a landmark global agreement—the Montreal Protocol on Substances that Deplete the Ozone Layer. The Protocol called for a freeze on production and use of halons at 1986 levels by mid-1989, and a reduction in CFC production by half over the next 10 years. It has since been amended four times to control additional substances and adjusted six times to accelerate schedules for phaseout. One of the most critical changes to the original agreement occurred in 1990 with creation of the Multilateral Fund for the Implementation of the Montreal Protocol (MLF). This fund is paid for by developed nations, and used to pay the agreed incremental costs for developing nations to switch to ozone-safe chemicals.

Since 1990, developed countries have contributed over $2.5 billion to support over 6,200 projects and activities in 148 countries implemented through agencies and by bilateral projects. A total of 446,173 ODP tons had already been phased out by the end of December 2009 (consumption of 249,494 ODP tons and the production of 196,679 ODP tons). To facilitate the phase-out, the MLF has funded ozone offices in 143 developing countries.

The Montreal Protocol provides a useful model and tool for other long-term environmental challenges such as climate change. The diplomats faced substantial difficulties that are now familiar in climate negotiations. Foremost was that the threat of ozone depletion was a global problem that affected every country and, hence, required global participation in the solution. Other difficulties were the genuine scientific uncertainty of the scale of harm and long-term nature of the environmental risks, sharply unequal regional and national contributions to the problem, potentially high costs to transition to new chemicals, and the unequal capacity among the countries required to bear the costs.

Despite these challenges, the Protocol secured global commitment to phase out chemical substances that were important to both consumers and businesses in the belief that reasonably priced alternatives would become available. The Protocol committed to providing financial support for all the incremental costs incurred by developing countries in achieving their phaseouts. Developed countries were called on to provide this financial support and to be first to undertake emission reductions. Developing countries were given a 10-year delay in meeting control requirements, and were provided financial support in meeting these control obligations.

In 2010, the Montreal Protocol became the only treaty on any subject to achieve universal ratification by every United Nations country in the world, which is a tribute to the global recognition and unity in actions to protect the Earth for future generations.

Trends in Stratospheric Ozone

The World Meteorological Organization (WMO) and UNEP regularly summarize worldwide scientific consensus on stratospheric ozone. The most recent reports are the *Scientific Assessment of Ozone Depletion: 2010, The Environmental Effects of Ozone Depletion and Its Interactions with Climate Change: 2010 Assessment,* and the *Technology and Economic Assessment: 2010.* Each report was prepared by a panel of experts from around the world. These assessments play a critical role in developing a common understanding of the scientific, economic, and environmental issues related to ozone depletion. They are relied upon extensively by policymakers in deciding what actions over time are required to modify the Montreal Protocol to respond to new information and reduce the risks of ozone depletion.

Previous assessments were published in 2006, 2002, 1998, 1994, 1991, and 1989. The experts have emphasized that, although natural sources put some chlorine and bromine into the lower atmosphere, human-induced bromine and chlorine compounds were causing significant polar ozone depletion. Moreover, the UNEP/WMO assessments have concluded that by 1980 most of the chlorine and bromine reaching the stratosphere came from human sources. From the early 1980s until the mid-1990s, measurements showed a continuing downtrend in global ozone amounts.

The Antarctic ozone hole is a persistent annual feature, with a level of ozone depletion that has been relatively stable since the early 1990s. Meteorological changes cause some year-to-year variation in ozone hole features. According to the UNEP/WMO 2006 Assessment, "Our basic understanding that anthropogenic ozone-depleting substances have been the principal cause of the ozone depletion over the past few decades has been strengthened."

The Antarctic ozone hole increased in size during the early 1990s, but at a slower rate than it had in the 1980s. At the turn of the century, the monthly total column values continued to be 40–50% less than pre-ozone hole conditions, and there was particularly extensive ozone loss in the lower stratosphere—the 7-to-12 mile (12-to-20 km) range. (Figure 35.)

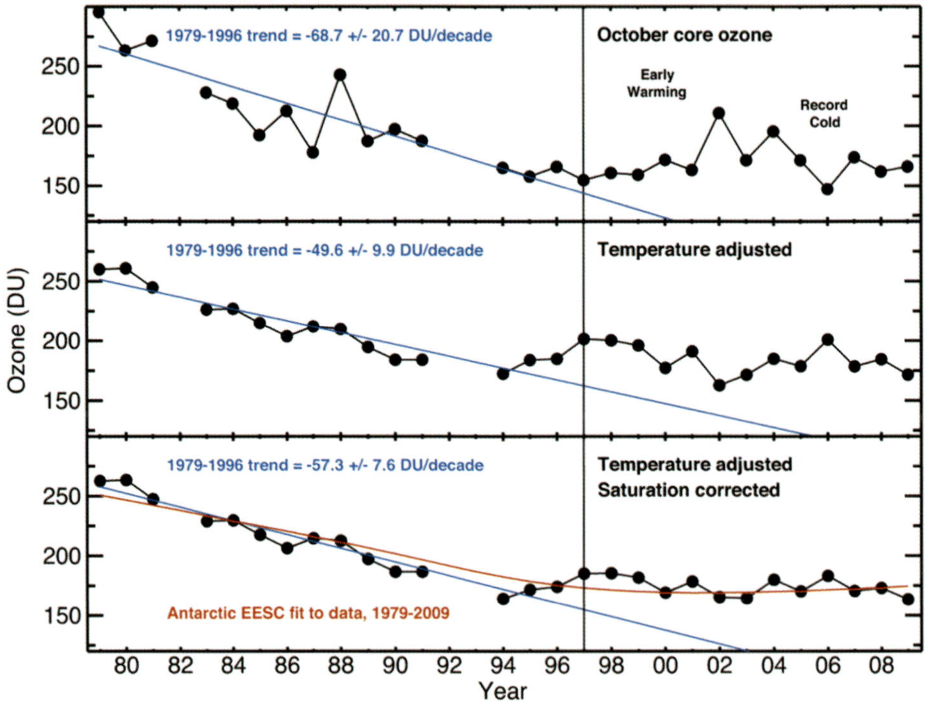

Figure 35. *Total column ozone in the core of the Antarctic vortex during October. Source: UNEP, 2010 Assessment of the Scientific Assessment Panel.*

The ozone layer above the Antarctic is not the only area affected. Significant ozone reductions occurred above the Arctic during the late winter and early spring in 7 out of 10 years in the 1990s, and 7 out of 9 years between 2000 and 2009.

When the ozone hole started appearing in the Antarctic in the 1980s, no similar hole appeared in the Arctic. Instead, late-winter/early-spring ozone depletion gradually eroded the normal high values of total ozone. After the winter of 1999–2000, NASA and a team of European scientists reported a loss of more than 60% of the ozone in a layer at about 11 miles above the Arctic. Because the Arctic has different weather patterns than the Antarctic, the Arctic ozone reductions were smaller than the reductions in the Antarctic ozone hole.

Ozone losses have also been observed in the mid-latitudes (35° to 60° North or South). In the mid-latitudes of the Northern Hemisphere, ozone losses were about 3% between 2002 and 2005 compared to the 1964–1980 average, according to UNEP/WMO, and larger in the spring. In the Southern Hemisphere, losses at the mid-latitudes were about 5.5% all year. Scientists have not observed any downtrend in total column ozone near the equator.

> *Our basic understanding that anthropogenic ozone-depleting substances have been the principal cause of the ozone depletion over the past few decades has been strengthened.*
>
> UNEP/WMO 2006 Assessment

Trends in Production and Use of Ozone-Depleting Compounds

Between 1986 and 2009, worldwide consumption of CFCs -11, -12, and -113 decreased by 97%, in most years substantially ahead of the control schedule required by the Montreal Protocol and its Amendments and Adjustments. Chemical manufacturers quickly commercialized new substances with low- or no-ODP, and product manufacturers who had come to depend on CFCs and other ODSs moved quickly to alternative processes and chemicals.

So far, at least 80% of ODSs that would have been used and emitted have been replaced with "not-in-kind" (NIK) technology, while 20% has been replaced with "in-kind" chemicals—mostly HFCs that are potent GHGs. Now, efforts are underway to replace these HFCs with new chemicals that have no impact on ozone and little impact on climate. Reported production of CFC and other ODS has continually declined since its peak in 1988, and in 2010 is less than 2% of peak use and emissions. All fluorocarbon manufacturers (including chemical companies in India and China) voluntarily report information annually.

Reductions in CFC use in industrialized countries have occurred in various ways, by:

- substituting not-in-kind hydrocarbons as aerosol propellants, as blowing agents for flexible foams, and as refrigerants in small refrigerators, refrigerated food-display cases, and window air conditioners;
- using NIK no-clean technologies in electronics manufacturing and assembly and NIK aqueous solvents in aerospace applications;
- substituting NIK alcohol and other solvents in place of CFC solvents in uses where aqueous and no-clean is inappropriate;
- recovering and recycling CFC refrigerants and recovering and destroying foam-blowing agents;
- increasing the use of HCFCs as transitional substitutes for CFC, which themselves are also being phased out; and
- increasing the use of HFCs that are ozone-safe, but are GHGs, for refrigeration and air conditioning and for minor applications such as specialty fire protection.

In 2009 and 2010, industry began a new transition to eliminate the HFCs that were used to replace the CFCs and HCFCs. The most conspicuous announcement was by General Motors, pledging to replace HFC-134a (GWP=1440) with HFO-1234yf (GWP=4) in car air conditioners. It is also significant that General Motors plans to increase the energy efficiency of its car air conditioners by at least 30%, which will reduce U.S. fuel use by 3 billion gallons a year if other automakers follow GM's leadership.

The 2010 UNEP/WMO Assessment concludes:

> The Montreal Protocol and its Amendments and Adjustments have successfully controlled the global production and consumption of ODSs over the last two decades, and the atmospheric abundances of nearly all major ODSs that were initially controlled are declining. Nevertheless, ozone depletion will continue for many more decades because several key ODSs last a long time in the atmosphere after emissions end.

The assessment also reports that bromine concentrations in the troposphere peaked around 1998 and have been in decline since, but it also warned that: "About half of the remaining methyl bromide consumption was for uses not controlled by the Montreal Protocol (quarantine and pre-shipment applications)," and noted that:

> Springtime Antarctic total column ozone losses (the ozone hole), first recognizable around 1980, continue to occur every year. Although the ozone losses exhibit year-to-year variations that are primarily driven by year-to-year changes in meteorology, October mean column ozone within the vortex has been about 40% below 1980 values for the past fifteen years. The average erythemal ("sunburning") UV measured at the South Pole between 1991 and 2010 was 50–85% larger than the estimated values for the years 1963–1980.

Scientists expect the Antarctic ozone hole to persist for decades, and no significant improvement is expected in the next 20 years. Antarctic ozone levels are projected to return to pre-1980 levels around 2060–2075.

There remain uncertainties in projecting levels of stratospheric ozone over the next century. For example, the IPCC points to uncertainties in: (1) the level of future consumption of ozone-depleting substances by developing countries; (2) projected levels of methane and nitrous oxide; and (3) projected climate change impacts on stratospheric temperatures and circulation.

Health Impacts of Increased UV-B Radiation

Ozone in the stratosphere protects the Earth from damaging amounts of UV radiation—a depleted ozone layer allows more of the Sun's damaging rays to reach the Earth's surface. Exposure to UV-B radiation can result in premature skin aging, increased incidences of skin cancer, damage to eyes, and suppression of the immune system.

Factors other than stratospheric ozone levels also affect the level of UV-B reaching the Earth's surface in a particular area. The primary factor is the angle of the sun's rays through the atmosphere. The more directly overhead the sun, the more UV-B reaches Earth's surface. Cloud cover, altitude, surface reflectivity, and aerosols also affect the amount of UV-B reaching the surface, and particles in the air such as smoke and dust block UV radiation.

According to UNEP/WMO, "Measurements from some stations in unpolluted locations indicate that UV irradiance (radiation levels) has been decreasing since the late 1990s, in accordance with observed ozone increases."

INCREASES IN SKIN CANCERS

Skin cancer is the most common type of cancer in the United States, with about a million new cases each year. UV radiation from the sun is the main cause of skin cancer, according to the Centers for Disease Control and

Prevention. The National Cancer Institute reports that 40–50% of Americans who live to age 65 will have some form of skin cancer at least once.

There are three primary types of skin cancers: basal cell, squamous cell, and malignant melanoma. Basal and squamous cell skin cancers are the two most common types. If detected early, these cancers are treatable. The third type, malignant melanoma, is far less common, but substantially more harmful.

Scientists agree that increased UV-B exposure increases the incidence of the milder basal and squamous cell skin cancers, and in 2002, UNEP reported on additional genetic information directly linking UV-B exposures to basal cell carcinoma in humans. Malignant melanoma is related to sun exposure in early life and to episodes of severe sunburn. UNEP reports that while much progress has been made in identifying genetic changes in melanoma cells, the precise mechanism of the progression to melanoma—and the role of UV radiation—"remain to be resolved."

> **UV INDEX**
>
> The UV Index is an internationally agreed-upon public health metric to inform citizens of the potential harmful effects to the skin of peak UV radiation on a particular day. A computer model is used to calculate the UV reaching the Earth, based on the forecasted ozone in the atmosphere at a particular location. The calculation is then weighted based on potential harmful effects to the skin of the specific UV wavelengths. It is then adjusted for forecasted cloud cover and elevation of the area. Overcast conditions can block out up to 70% of UV radiation.

SUPPRESSION OF THE HUMAN IMMUNE RESPONSE SYSTEM AND INFECTIONS

UV radiation weakens the ability of the immune system to respond to some infectious agents and some cancers. Exposure to UV radiation can trigger two different types of herpes infections (cold sores and shingles), according to UNEP. Exposure to UV radiation can also reduce the effectiveness of vaccinations, regardless of whether the exposure occurred before or after the immunization. There is still much that is not known about the full implications of effects of increased exposure to UV radiation on the immune system.

DAMAGE TO HUMAN EYESIGHT

UV radiation can damage the cornea and conjunctiva (mucus membrane covering the eye), the lens, and the retina. Acute exposure to UV can cause photokeratosis, or "snow blindness," similar to sunburn of the cornea and conjunctiva. Long-term exposure to UV is associated with the development of cataracts, which cloud the lens of the eye, limiting vision, and which, if not treated, can cause blindness.

INTERACTIONS BETWEEN CLIMATE CHANGE AND OZONE DEPLETION

Scientists have long understood that in addition to destroying ozone in the stratosphere, CFCs also act in a similar manner to carbon dioxide and are greenhouse gases. In fact, in part because of their long atmospheric lifetimes, CFCs are 5,000 to 8,500 times more potent than carbon dioxide in contributing to global warming.

Actions under the Montreal Protocol thus made large contributions toward reducing global greenhouse gas emissions (Figure 36). In 2010, the decrease of annual ODS emissions under the Montreal Protocol is estimated to be about 10 gigatons of avoided CO_2-equivalent emissions per year, which is about five times larger than the annual emissions reduction target for the first commitment period (2008–2012) of the Kyoto Protocol.

The story is somewhat more complicated, however, because when CFCs and other ODSs were phased out, some were replaced with HCFCs (which have low ODP and generally lower GWP) and with HFCs and other compounds that are ozone-safe but are also greenhouse gases (though generally less potent than the CFCs and HCFCs they replace). In fact, the HFCs currently used as ODS replacements contribute about 0.4 gigatons of CO_2-equivalent per year to total global emissions, while the currently used HCFCs contribute about 0.7 gigatons. CO_2-equivalent emissions of HFCs are increasing by about 8% per year, and without fast action under the Montreal Protocol to control HFCs, this rate will continue to grow.

The 2010 UNEP Assessment warned that:

> As the majority of ODSs have been phased out, demand for hydrochlorofluorocarbon (HCFC) and hydrofluorocarbon (HFC) substitutes for the substances controlled under the Montreal Protocol has increased; these are also greenhouse gases....For example, abundances of HFC-134a, the most abundant HFC, have been increasing by about 10% per year in recent years. Abundances of other HFCs, including HFC-125, -143a, -32, and -152a, have also

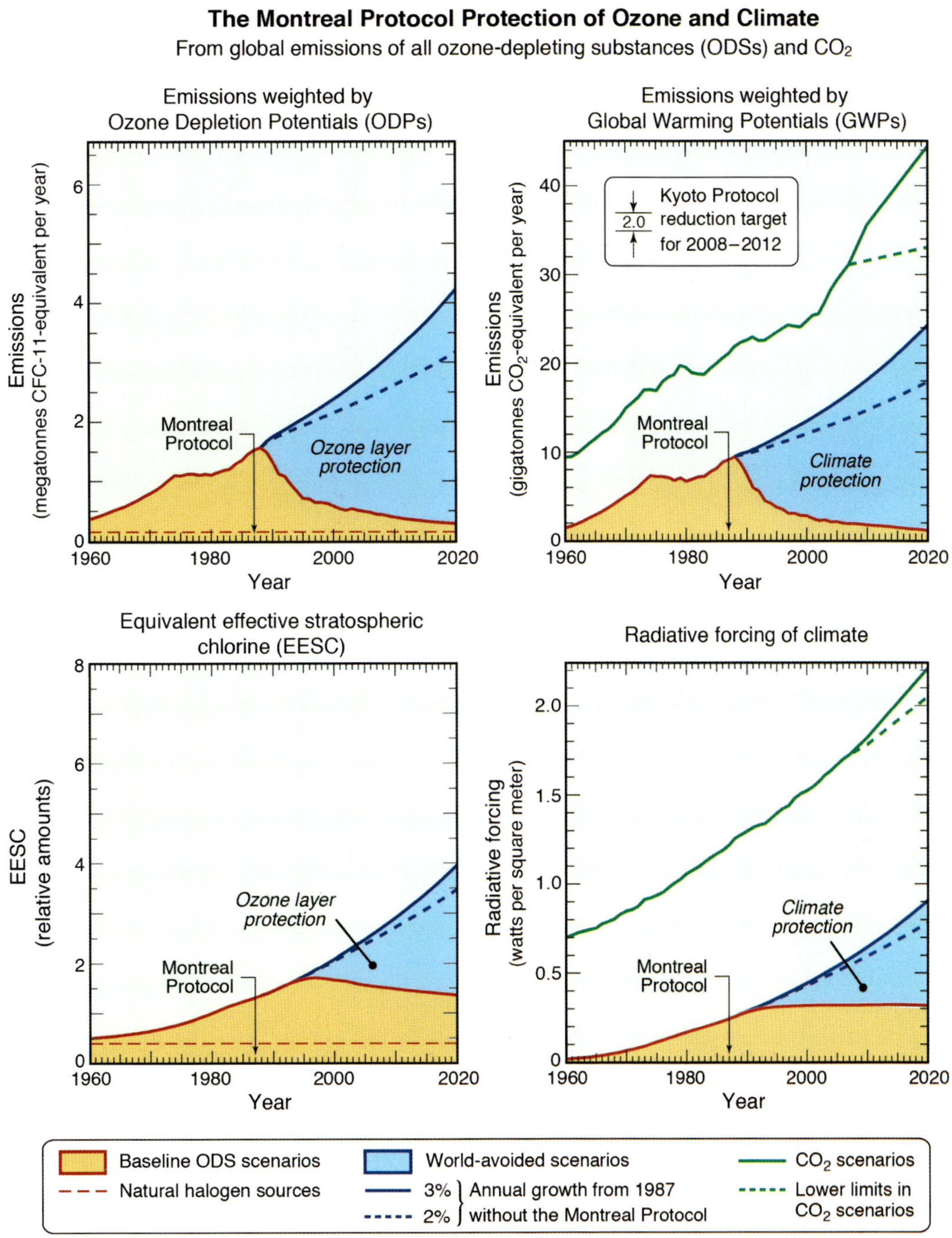

Figure 36: *Montreal Protocol protection of ozone and climate. The provisions of the Montreal Protocol have substantially reduced ozone-depleting substances (ODSs) in the atmosphere. This has protected the ozone layer and also reduced the potential for climate change because ODSs are greenhouse gases. The scenarios and comparisons shown here demonstrate this dual benefit of the Montreal Protocol. Baseline scenarios for ODS emissions include all principal gases weighted by their Ozone Depletion Potentials (ODPs) or Global Warming Potentials (GWPs) (top panels). With these weightings, emissions are expressed as CFC-11-equivalent or CO_2-equivalent mass per year. The lower panels show EESC and radiative forcing of climate as derived from the respective ODP- and GWP-weighted scenarios. The world-avoided emission scenarios assume ODS emission growth of 2 or 3% per year beyond 1987 abundances. Shown for reference are the emissions and radiative forcing of CO_2, and the emissions reduction target of the first commitment period of the Kyoto Protocol. The contributions of natural halogen source gases are shown in the ODP-weighted and EESC scenarios (red dashed lines) and are negligible in the GWP-weighted and radiative forcing scenarios. The magnitude of the dual benefit has increased since about 1987 as shown by differences between the world-avoided and baseline scenarios (blue shaded regions in each panel). For completeness, these differences can be adjusted by offsets due to additional ozone depletion and HFC emissions (see text). (A megatonne = 1 billion (109) kilograms. A gigatonne = 1 trillion (1012) kilograms.) Source: NOAA et al., Twenty Questions and Answers About the Ozone Layer: 2010 Update.*

been increasing. Regional studies suggest significant HFC emissions from Europe, Asia, and North America.

In response to concern over the climate impacts of HFCs used to replace ODSs, two similar proposals have been put forward to amend the Montreal Protocol to phase down the use of HFCs. The first proposal is from the Federated States of Micronesia and the second from Mexico, Canada, and the United States. At first, many developed countries hoped that the 2009 Copenhagen agreement would bring controls to HFCs and that action under Montreal would be unnecessary; and many developing nations hoped that financing under the Clean Development Mechanism (CDM) of the Kyoto Protocol would be more profitable than under the Montreal Protocol.

However, Copenhagen failed to reach a strong agreement, and action on HFCs under the CDM is stalled by controversy and lack of funds. Countries are considering this action under the Montreal Protocol because HFCs were introduced as substitutes for CFCs and the Parties to the Protocol have a good understanding of the sectors and alternative technologies. In addition the Montreal Protocol has well established and highly regarded institutions already in place and ready to act quickly, skillfully and flexibly to control HFCs. The parties to the Montreal Protocol are expected to take up this issue again at their meetings in 2011.

CHAPTER 11: WORKING WITH SCIENTISTS AND SCIENTIFIC NEWS SOURCES

Introduction

Science, along with economics and political and policy considerations, is a critical component in the climate change/global warming story. For many journalists, covering science responsibly presents particular challenges. As science stories go, climate change is particularly difficult and complex—and far more politically charged than many other science-based issues. It is also a story that is certain to be around for years and decades to come.

The issue is loaded with potential journalistic traps and pitfalls. This chapter suggests some ways of overcoming some of them.

The most authoritative sources of climate science information are scientists doing "good science" and reputable peer-reviewed professional journals describing their work. But those aren't the only reliable sources, and many scientists are reluctant to talk with journalists and are often difficult to understand when they do. Understanding the language, customs, and ground rules of responsible science is essential to encouraging a sharing of scientists' valuable insights. There is no substitute for doing your homework before you talk with scientists.

Pitfalls? Many science stories in the popular press simply turn out to be wrong, either because they were inaccurately reported or because the underlying science was flawed, bogus, or simply politicized. In recent years, the media have reported enthusiastically on all kinds of questionable scientific advances, such as cold fusion, cloned human beings, or any number of "cancer cures."

On climate change, it's worth remembering that in the late 1970s the American news media were abuzz with talk of a "coming Ice Age." Some reports spoke of a "snowblitz" theory in which advancing glaciers engulf civilization not inch-by-inch over the millennium but in one sudden surge. Climate contrarians repeatedly point back to these erroneous news reports to support their flawed idea that scientists don't know what they're talking about. In reality, although a few scientific papers did postulate global cooling, the scientific community never reached a consensus on this prediction.

But there's help out there. Reporters can avoid going too far wrong in reporting on climate science by rigorously adhering to a few simple principles:

- *Trust only* ***peer-reviewed*** *science.* Beware giving too much credence to scientific studies not published in recognized standard journals of the appropriate scientific disciplines. Be skeptical of press conferences announcing dramatic scientific findings and of political scientists in the guise of hard scientists.
- *Build stories on* ***consensus*** *science,* while recognizing that many areas of eventual scientific consensus originate as minority viewpoints. Even in controversial issues such as climate science, there are considerable areas of widespread scientific agreement (for instance, no one seriously questions that there is a greenhouse effect). Consensus science generally is found in reports and assessments done by scientific organizations, often involving scores or even hundreds of researchers. Beware of the deadline-driven frenzy to rush to air or print dramatic scientific findings based on a single study. Take the time to understand the existing context of scientific research and findings in which new findings must be evaluated.
- *Beware when hearing scientific information from non-scientists.* Be skeptical of scientific claims and interpretations from groups with an economic or ideological stake in the climate issue, whether from an industry or an environmentalist perspective. Know who funds the research and who pays the salaries, and use that information as one factor in evaluating the weight of a scientist's credibility.
- *Keep* ***uncertainty*** *in perspective.* It's a critical part of science, and the most authoritative scientists won't shy away from it in considering their own work. Acknowledging and quantifying uncertainty are often hallmarks of the most responsible scientific studies, and denial of it can be a warning sign. Some on all sides in the climate change/global warming debate try to manipulate the uncertainty issue to their own advantage, shifting the burden of proof to the opposing perspective. That said, uncertainty does not in itself imply lack of knowledge.
- *Don't assume that "balance" is an adequate surrogate for accuracy, fairness, and thoroughness.* Don't

assume any "truth" is indelible and not subject to further and future rethinking. The "he said-she said" formula and the "On the one hand, this/on the other hand, that" template may not meet the public's essential information needs, particularly on a scientific subject. Balancing authoritative assessments with opposing, but unreliable, viewpoints can promote misinformation.

COVERING THE SCIENCE OF CLIMATE CHANGE

Advocates on all sides of the climate change issue assert that they are backed up by "good science." Reporters often lack training or a detailed grasp of the underlying physical sciences to evaluate their claims. They may or may not have started out as science writers, but increasingly, they find science an essential ingredient in effective environmental journalism. In the field of climate change, the number of specific scientific disciplines involved compounds the problem.

Covering science stories for the public often means starting at square one. Among the general public, science literacy is a fundamental weakness. According to longtime and well-respected science writer Jon Franklin, writing in the Fall 2002 issue of *Nieman Reports* on "The Extraordinary Adventure That is Science Writing":

> It's not so much that science is more difficult—you want complex, look at the rules governing baseball. But most Americans have a context for baseball, and they have practically none for science. That means you have to give a whole lot of backstory. That takes time and space, and most of all it takes experience.

THE LANGUAGE OF SCIENCE

Journalists' and scientists' information goals and methods often overlap, according to former *Washington Post* science reporter Boyce Rensberger, now at the Massachusetts Institute of Technology (MIT). "Both seek truth and want to make it known. Both devote considerable energy to guard against being misled. Both observe a discipline of verifying information."

The best scientists and journalists also seek findings that will stand up to review and scrutiny, but there also are some important differences. Scientists and journalists often differ in their approach to telling the story. Journalists are in the business of providing news and information and, increasingly in the culture of many modern newsrooms, doing so in a way that also provides entertainment.

Scientists seek scholarly communication and public education about their areas of interest. They often strive for cautious, precise, qualified, and complex explanations for things, with clear warnings against misinterpretation or overinterpretation. Unlike journalists, they often put their major findings at the end, rather than at the beginning, of their narratives. They often are disappointed when journalists appear not to understand and appreciate, fully, where they are coming from, or when editors and reporters seek a more definitive answer than they may be comfortable providing.

Scientists have their own language for talking with each other. One researcher proposes a hypothesis, another devises an experiment to test it, a third tries to duplicate the experiment and sometimes gets different results. Science advances because ideas are constantly being tested, discussed, and revised. It's not at all unusual for different scientists to look at the same data and come up with different conclusions. Scientists talk to each other in certain formal channels. They publish articles in often narrowly focused, peer-reviewed scientific journals, deliver papers at professional meetings, and sit on committees that put together reports.

Scientists use technical language partly as a convenient shorthand, and partly as a badge of their expertise and status, write Michael R. Greenberg, Peter M. Sandman, and David B. Sachsman in their risk communication guidance. "When scientists and engineers talk to each other, they often mix the jargon with a good deal of less formal, less technical language. But on formal occasions they naturally return to the precision of a highly controlled vocabulary. An interaction with a reporter is likely to be treated as a formal occasion; it is going 'on the record' on matters of professional and public importance."

THE ROLE OF PEER REVIEW AND CONSENSUS

Good scientists go through a process of replicating results, publication, review, discussion, and revision, bringing other workers in their field into the process of interpreting results. This ritual happens in scientific journals, conferences, and committee meetings. One term for it is **peer review**. Without it, the public tends to get more stories like the miraculous but implausible "cold fusion" breakthrough, which turned out to be far less than early accounts suggested.

Eventually, if a conclusion holds up in the face of peer review, after enough meetings and articles, it may win a consensus within the scientific community—or a near-consensus, since virtually nothing in science is ever finally settled for all time.

THE SCIENTIST'S PERSPECTIVE

Most scientists resist some of the media tendency to present their work as more final, more certain, or conclusive than it really is. As in good writing, shades of subtle meaning are extremely important to responsible scientists. An epidemiologist may plead with a reporter not to write that he "suspects" a certain virus as the cause of an epidemic, or that he is working on the "hypothesis" that it is the cause. The right way to say it, the epidemiologist may insist, is that it "has yet to be ruled out."

In this era of tabloid TV, even experienced headline writers, editors, and reporters are often posing questions which may be very interesting, controversial, or provocative—but also may be very hard to answer with scientific precision.

Think of headlines along the lines of: "Global Warming, Threat or Hype?" . . . "Air pollution: Who's to Blame?" . . . "Climate Change: Who Are the Winners and Losers?" . . . or "Energy Conservation: Success or Failure?" Often, there are no up-down, either-or answers for such questions.

Such questions may involve politics, social values, undeclared motives, or governmental policies. And they may be the kind of questions that scientists are neither able to answer nor willing—speaking as scientists—to answer. When scientists refuse to answer or appear to duck these questions, many journalists get frustrated and start pressing hard, which can be counterproductive in terms of keeping communications flowing between scientists and the news media.

Scientists will often bristle at attempts by the media to personalize their work. They have been trained to keep themselves separate from their research. Personal questions, such as what they or their families would do under the circumstances, are not welcome, although to many media representatives such questions are entirely appropriate and, in many cases, necessary.

Scientists who normally publish their work in technical journals often see coverage in the popular press as a no-win proposition. If the coverage turns out to be misleading or inaccurate, they may fear that their reputation among their colleagues is threatened. Yet they also recognize that continued funding often depends on broader public awareness of their research. But given the general absence of institutional incentives in the scientific community to work with media, they frequently feel that even "fair" coverage can provide minimal benefits among their scientific peers.

Often, scientists fear that a reporter will misquote, simplify, or distort their work in a way that misrepresents it before their professional colleagues or misinforms the public, causing damage that the scientist will then have to try to undo. Even defenders of the media's responsibilities and prerogatives acknowledge that this happens all too often.

There are of course some scientists who actively seek out publicity through media coverage and are more than eager to talk with reporters and to be quoted. They may simply want recognition for their work, like everybody else. Also, research funding is scarce and difficult to obtain, and many scientists hope media attention will help them win more financial support for their work.

At the same time, most scientists are unaccustomed to dealing with the media and find it hard to explain their work in simple terms. "Scientists have important roles to play in getting the news right," says Cornelia Dean, *New York Times* science reporter. "But they often are reluctant participants." When speaking with scientists about interviews with the press, Dean advises them to prepare as they would for a professional presentation. They should determine which points are most important and how to cover them clearly and simply, and use visuals like graphs and charts to add clarity. And they should be encouraged to try to address reporters' questions.

HOW TO MAKE THE INTERVIEW A SUCCESS

Advance preparation, listening attentively, and establishing a level of trust are at the heart of virtually all successful interviewing. But some aspects are particularly important for reporting on climate change. Among these:

Do your homework. "You've got to know something about the subject before you go chase it," says Ben Patrusky, Executive Director of the Council for the Advancement of Science Writing, referring to the issue of global climate change. "Without some basic knowledge, you shouldn't be covering this at all." That does not mean the field is restricted to reporter-specialists. It does mean doing "a little homework."

"Know what the scientist's field is, and a little about the field," advises longtime science writer David Perlman, of the *San Francisco Chronicle*. "Know what the scientist's focus is. No reporter should attempt to interview a scientist without some familiarity with the field in which the scientist is working."

Have questions prepared beforehand, but always be willing to stray from them. "The first question I ask in the actual interview session is critical," says *New York Times* writer Claudia Dreifus. "It sets the tone for everything that will happen subsequently. It shows that I'm serious, and that I've done my preparation, and thought a lot about the subject and his or her work."

"Questions should not be so broad that they become difficult for the scientist to answer," adds Perlman.

Double-check for accuracy. Reporters should not be surprised to find scientists asking to see a copy of their article before publication. This is a common courtesy between scientists in the culture of science. Some scientists, however, may not be aware that in the culture of journalism, it presents a threat to the journalist's independence. Many news organizations ban such prior reviews. On the other hand, some well-respected and established science writers use the practice with discretion to help ensure accuracy, and they offer no apologies to reporters finding it a threat to journalistic independence.

Patrusky says there is nothing wrong with calling a scientist back to read particular passages in a story, check particular facts, or verify particular direct quotes. But he says a journalist clearly is not obligated to show a scientist-source a story prior to publication. And certainly, he says, the journalist is not obligated to let scientists judge the emphasis and interpretation of the story. "They're just as biased as the journalist in their way," Patrusky says.

Ask for opinions on broader issues, but don't be surprised if they are not forthcoming. Scientists by virtue of their training can be reluctant to share their views on policy issues, or to speculate beyond the narrow vertical confines of their field of specialty. Nonetheless, their opinions about public policy issues pertinent to their fields may be quite valuable. Even if those are personal and non-scientific judgments, scientists are often better informed than the average person. When you want policy opinions, Perlman suggests, it may be fruitful to phrase questions that specifically require policy opinions, and to push the scientist a little if you don't get an answer.

But Perlman worries more about scientists who are uninhibited with reporters than about those who are restrained: "The worst scientists are the ones who give you a lot of policy decisions that they are not qualified to give."

Radio presents unique challenges, because stories often compete with a host of potential interruptions. One digression, or lapse into the esoteric, can be the kiss of death. All the same, Christopher Joyce, writing in *Nieman Reports*, argued that the "doing of science is rich territory for radio, since it's full of sound, if not fury." He urges science reporters to exploit the medium to report the "process" of science, rather than just the results. "When you interview scientists," he urges, realize that you don't have to 'dumb down' the concept. Complicated and abstract ideas are fine; just eliminate complicated and abstract language."

DEALING WITH UNCERTAINTY

Uncertainty about what scientists are saying, and what should be done about it, is at the core of the public policy debate over climate change. Yet journalism thrives on "facts," and editors often seek facts that are absolute and unequivocal. In practice, this approach can lead to reporters' having to defend their writing against editors and fact-checkers seeking a "truth" that may or may not exist.

Boyce Rensberger, since 1998 director of MIT's Knight Science Journalism Fellowships, explains the difference between textbook science, with its well-established facts, and what scientists actually do for a living. Scientific findings in his view are always hedged in uncertainty. "Scientists don't get grants to discover what is already known," he explains. "Uncertainty is a sign of honest science and reveals a need for further research before reaching a conclusion." He cautions against reading too much into the results of a single experiment or study. "The pace of science, despite the hype, is usually slow, not fast. Breakthroughs are never the result of one experiment."

Figure 37. *Boyce Rensberger, Director of the Knight Science of Journalism Fellowships program at the Massachusetts Institute of Technology. "Experienced science writers try to keep the sense of uncertainty in the copy," says Rensberger. "But too often, editors instinctively strike out the caveats that, in their minds, weaken the story. Headline writers further prune perspective and judgment." Source: Graham Ramsay.*

"Experienced science writers try to keep the sense of uncertainty in the copy," says Rensberger. "But too often, editors instinctively strike out the caveats that, in their minds, weaken the story. Headline writers further prune perspective and judgment."

"When it comes to almost anything we say," reports Dr. Arnold Relman, a former editor-in-chief of the *New England Journal of Medicine*, "you, the reporter, must realize—and must help the public understand—that we are almost always dealing with an element of uncertainty. Most scientific information is of a probable nature, and we are only talking about probabilities, not certainty.

What we are concluding is the best we can do, our best opinion at the moment, and things may be updated in the future."

Douglas Starr, Co-Director of the Science and Medical Journalism Program at Boston University, encourages his students to pursue uncertainty and not shy away from it: "Areas of uncertainty represent the cutting edge of science and provide insights into the scientific debate," advises Starr. "Part of what makes the global warming debate so compelling, for example, is what society should do given the uncertainty about the dimensions of the problems."

Figure 38. *Douglas Starr, at the Knight Center for Science and Medical Journalism at Boston University: "On the issue of global warming, should we give as much weight to the handful of naysayers known to be supported by the fossil fuel industry as we give to the more than 2,000 climatologists from 120 countries represented on the Intergovernmental Panel on Climate Change?"*

New York Times science writer Dean puts it simply: "Science gyrates," she cautioned a meeting of New England science editors and reporters at a *Providence Journal* conference in 2003.

At the same time, reporters have to strike a delicate balance between being honest about the persistent uncertainties, while at the same time conveying the reality that we do know quite a bit—and in the case of climate change, we know enough to act.

BALANCE—OR ACCURACY?

Daily journalists with too little time and space have long fallen back on the device of presenting two extreme, opposing viewpoints as a way of achieving "balance." But reporters with time for one more phone call might do well to ask themselves whether "balance" is always an adequate substitute for accuracy and fairness, for completeness, or for real understanding. While striving for "balance" may be helpful when reporting on areas involving conflicting opinions and perspectives, it may be less useful when reporting on hard-and-fast scientific findings—even those involving uncertainties.

As a practical matter, many journalists and news organizations serving broad audiences don't often write about pure science stories; usually the story is pegged to some government action, finding, decision, or controversy. Consequently, one method many reporters use to assemble the story is the "insider-outsider" model. First, they quote the government official or government scientist, then some person or persons outside government, working to evaluate government policy or science. Special-interest lobbies and think tanks often add nuances, details, and "color" to such stories.

But reporters need to take care in weighing these opposing views, not just presenting them at face value. "On the issue of global warming," Boston University's Starr cautions, "should we give as much weight to the handful of naysayers known to be supported by the fossil fuel industry as we give to the more than 2,000 climatologists from 120 countries represented on the Intergovernmental Panel on Climate Change?" Clearly, he thinks not.

Speaking at an April 2003 *Providence Journal* science editors/reporters meeting, the National Science Foundation's Assistant Director for Geosciences Margaret Leinen pointed out that "reporters generally consider everyone's opinion equally." She cautioned that as a result the media may give too little emphasis to "the 96% of facts that we all agree on," and too much to remaining issues still contested among responsible scientists.

"The more uncertain the issue is, the more people you have to talk to," says Patrusky. "The best instruments that science writers have are telephones," he adds. "I'm a student of 'check it out, check it out, check it out.' And the more uncertain it is, check it out with more people."

But how can you be sure you've reached the right people? Starr stresses the importance of stating where the consensus of scientific opinion lies, and of revealing the sources of support of the various opinionmakers. This contextual reporting will, he has written in *Nieman Reports*, avoid the "he says, she says" dilemma of traditional reporting and will help provide increased depth of coverage.

Reporters may be on much safer ground relying on consensus science than on science that the scientific "mainstream" holds to be well established. The corollary to that argument, of course, is that the minority often turns out over time to have been "right." Reporters need to be mindful of that reality, while not overplaying a fringe or "wacky" perspective.

As described earlier, scientists reach consensus on certain topics through their professional organizations, meetings, collective authorship, peer review, and similar mechanisms. In the climate area, groups like the Intergovernmental Panel on Climate Change, the World Meteorological Organization, and the National Academy of Sciences have issued reports summarizing the knowledge most scientists agree on. These are a handy way to locate the most responsible and authoritative thinking.

The problem with the point/counterpoint, for/against approach that is popular in much journalism is that readers or viewers may be left to average out the various viewpoints and come to their own conclusions. Too often the "point and counterpoint" average out to nothing at all—they just cancel each other out. "That doesn't advance people's understanding very much," says Perlman.

According to a group of Rutgers University authors in their manual, *Risk Communication for Environmental News Sources*: "The scientific establishment, like any establishment, sometimes turns out systematically wrong, but the scientific majority usually turns out closer to the mark than the mavericks. It is deceptive to present such opposition positions as if they had equal support."

DEALING WITH COMPLEXITY

For mass-media environmental reporters, adequate airtime and print space for in-depth coverage of complex scientific issues is an increasingly rare commodity. Their editors too often would have them shorten and simplify their stories; time for boning up on the subject is often at a premium.

Starr agrees that clarification is essential, but he cautions that ignoring complexity can lead to a partial, and at times misleading, story. "Such was the case last winter, when scientists reported that certain parts of the Antarctic ice sheet were thickening," Starr wrote in the 2002 *Nieman Reports* issue focusing on science journalism. "As journalist Keay Davidson, then of the *San Francisco Chronicle*, pointed out, some newspapers simplistically editorialized that the findings cast doubt on the theory of global warming. Actually, the findings shed light on the incredibly complex movement of polar ice sheets, including the likelihood that global warming will produce unstable weather patterns."

TIPS FOR EVALUATING SCIENTIFIC STUDIES

Reporters often face demanding deadlines, but they need to try to approach new studies and findings with questions such as the following:

- Was the study published in a recognized journal or presented at a recognized professional meeting?
- Has it been peer-reviewed by disinterested parties?
- Are the reviewers known, and if so, who were they? (Peer reviews often are conducted anonymously under guidelines established by individual journals.)

Not all science journals are created equal. One veteran science editor says the first thing you want know about a journal is whether it is **peer-reviewed**—with a submitted article reviewed by a panel of scientists working in the field it pertains to, then defended and revised in light of their comments. Often these are published by university-affiliated research institutes and professional associations.

Professional meetings and committee reports are also valuable sources, because these channels involve groups of scientists involved in dialogue and reflecting different opinions. Journalists covering them are less likely to fall victim to bad information. A reporter covering a source or event outside these channels should use additional caution to ensure the integrity of the reporting.

Some science findings are published as commercial ventures in "trade" or "industry" publishing. Naturally, peer-reviewed journals tend to be more disinterested and reliable.

A scientist or group of scientists holding a press conference to announce results before they have been published in journals or discussed at professional meetings may have good reasons for doing so; but they may also be handing the media another "cold fusion" story, which evaporates under scrutiny. The reporter who simply takes the handout may get on page one. But the reporter who seasons it with skepticism may also get on page one—and do the audience the service of better informing them and clarifying the subject.

Similar care is called for in other media events involving science, particularly given the inevitable politicization of science surrounding important public policy issues such as global climate change. Scientists testify at congressional hearings. They appear on talk shows and news-interview shows. Scientists may be employed or retained by a lobbying or interest group, and these groups may issue their own "reports." Some even take to the lecture circuit, talking on topics outside their specific areas of training and expertise.

Is the entire study available for reading and review? Fact sheets and abstracts, which usually summarize the findings and conclusions, are a good place to start, and may be all that is readily available to reporters on short deadline. But they seldom tell the whole story, and may in fact be misleading. ("I've never seen a factual fact sheet," some journalists are fond of saying. Keep in mind the refrain associated with sixties TV detective Sergeant Joe Friday: "Just the facts, all the facts, and nothing but the facts." Can you really say that many "fact sheets" you have seen meet that standard?)

For reporters who find the technical jargon and sheer length of journal articles daunting, one option is to get them "translated." An author who writes in equations for his professional colleagues may be able to explain the article in lay terms over the phone. The journal editor who published it may also be a good resource here. A credible scientist at the local university may also help.

Are the results credible? Do the results follow logically from the methods? Do the conclusions make sense? Greenberg, Sandman, and Sachsman have authored a number of guidebooks for environmental journalists and news sources. They say credible results score high for statistical significance, have been successfully replicated with similar results, have stood up under reevaluation by difference kinds of analysis, and honestly and openly describe data weaknesses, which inevitably exist.

Could the association claimed have resulted merely by chance? How have the investigators arrived at their finding of statistical significance?

Statistics is a branch of mathematics developed to translate certainty and uncertainty into numbers. Some understanding of statistics is essential to any understanding of climate change. What does it mean when scientists say something is **statistically significant**? It is easy to learn some basic concepts in statistics without understanding the difficult math. Two books that can help reporters overcome these challenges are the late science writer Victor Cohn's *News & Numbers*, and Phillip Meyer's *The New Precision Journalism*.

What is really known at this point? And what is unknown? Have the study authors acknowledged the limitations of the data? Do the authors frankly discuss possible flaws in their study? Do the data fully support the conclusions?

What are the broader implications of this particular study? Douglas Starr, of Boston University, teaches his students to use newsworthy journal articles as a point of departure for interviews, and other research to reveal the broader currents and trends. Starr advises his science journalism students to ask contextual questions about how the particular study compares with others in the field. He encourages reporters to ask questions such as: What similar studies have come before? How is this one different? How does this study add to or contradict the existing body of scientific opinion?

Because articles in science journals tend to focus on very specific questions, with tightly defined boundaries, the hardest part of understanding a journal article may be reading between the lines. While an article may confine itself to a specific question, its broader importance may lie in what it responds to—what other findings it seems to contradict, or what theory or hypothesis it confirms or undermines. In a lot of cases, reporters can get this kind of context only by doing a lot of homework, and strengthening their background understanding of issues such as climate change.

Do mainstream experts in the researcher's field of expertise generally concur with the conclusions and the processes by which they were reached? Except on noncontroversial items, says Perlman, "You almost never try to do a one-scientist story, because you always try to solicit some comment." But the reporter has to use some ingenuity to find people who are working in the same field or related fields. It is perfectly legitimate, he says, to ask a scientist you have called: "Who else is working in your field who may agree or disagree?"

Has the researcher or the organization championing the research previously taken positions on the issues being addressed? "Even the 'neutral expert' isn't as neutral as the media (or the expert) would like to believe," caution Greenberg, Sandman, and Sachsman in their guide for environmental news sources, *The Environmental News Source: Informing the Media during An Environmental Crisis*. "Those who don't have a vested interest in their funding source or their policy perspective may still have a vested interest in prior findings: when scientists look at a new development, they see it through the prism of their own work, preferring to see themselves confirmed rather than disconfirmed."

What groups or interests supported this study or the researcher's work in the past? The old adage, "follow the money" is good to keep in mind, however painful many news sources may find it. Reporters are advised to always ask who funded the study. They should not be surprised to find, more often than not, that study findings support the sponsor's agenda.

"We need to remember that opinions, connections, and sources of financing are not necessarily indictments," Cohn has cautioned. "But, given the topic, the public may have a right to know them. And there are prejudice peddlers."

"The scientists doing the studies, finally, are just people," observes Cohn. "A few, we have learned, can be dishonest. Many more are subject to biases, prejudices, economic interests, and political leanings that can color judgment."

In the end, of course, there's no journalistic "silver bullet" for ferreting out fact from fiction in reporting on climate change and global warming science. A sound prescription—one common to virtually all serious journalism: Take the time to understand and know the subject and the "players." Do your homework. Be skeptical, but not cynical. Choose your "experts"—and your words—carefully. In other words—good journalism.

CHAPTER 12: QUESTIONS NEEDING BETTER ANSWERS

A fundamental popular misconception about science is that it offers certainty on most things. But residual uncertainty, doubt, and healthy skepticism are fundamental aspects of science.

The press and the public too often fall prey to efforts to exploit such uncertainty by both sides in a given policy debate. The doomsayers call for action to avert a possible catastrophic outcome—while the contrarians call for inaction because the certainty of catastrophe has not been proved beyond any doubt. Both sides in the debate tend to exaggerate the certainties or the uncertainties.

Scientists working on climate have vastly improved their understanding in the last several decades. Many aspects of the climate system are quantitatively understood much better than they were a decade ago.

At the same time, a deeper understanding of the climate system has also heightened appreciation for its complexity and vast scales in time and space. Decades of study have resulted in better experiments, instruments, measurements, data protocols, and more sophisticated models representing more in-depth knowledge of physical processes. However, even as precision has grown in some areas, unknown or uncertain aspects have become even more obvious.

Filling in the remaining blanks will not be easy. The effort to do so will be as much organizational and economic as it will be scientific. Many of the important advances in climate knowledge have come from large-scale international cooperative efforts lasting perhaps a decade—the Joint Global Ocean Flux Study (JGOFS) is just one of many examples (http://ijgofs.whoi.edu). Two keys to improving understanding of climate are: the need for global-scale coverage in measurements, and consistency of measurements on timescales of decades and ultimately centuries.

Because of the complexity and vastness of the climate system and its interactions, uncertainties remain in a number of specific areas. Among them are the fate of land-based ice sheets in a warmer world; the roles of clouds, aerosols, and soils; the global carbon budget; the dynamics of atmospheric chemistry; ocean currents; ocean chemistry; changes in extreme weather; local impacts; and future human behavior.

However, it's important to emphasize that these uncertainties do not necessarily preclude policy actions. It's difficult to provide an example of any policy action that has been taken with absolutely complete information or without some inherent uncertainty in the outcome. And yet decisionmakers take action on a daily basis in the face of uncertainty. The climate change issue, therefore, is really one of reducing the **risk** of future impacts, and we should not be fooled into inaction by the remaining uncertainties in climate science.

Land Ice

As discussed in detail in Chapter 6: Sea-Level Rise, land-based ice is potentially a large—even primary—driver of eventual sea-level rise. Contributions to sea-level rise from the melting of mountain glaciers and other ice and snow on land is well understood. The big question here is how ice **dynamics** will change under a warmer climate. In other words, how much and how fast will large, land-based ice sheets break apart and slide directly into the sea?

Unfortunately, the response of land ice to global warming is not well understood and difficult to predict. The AR4 essentially punted on this question: the experts concluded that not enough was known about the dynamics of land ice to make a prediction of sea-level rise in 2100. Research on this question has continued

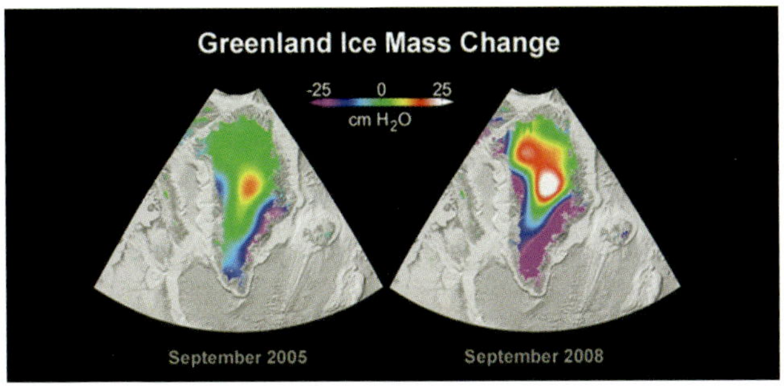

Figure 39. *Satellite measurements of ice-mass loss in Greenland, comparing September 2005 to September 2008. More recent data indicate that ice-mass loss has spread to the northwest coast, in addition to more severe losses in the southern region. Purple areas indicate the greatest rate of ice-mass loss. Source: JPL/NASA.*

since the publication of the AR4, and estimates range from 0.8 to 2.0 m of sea-level rise by 2100, but further questions remain.

Figure 40. *Source: Artville.*

Clouds remain one of the biggest unknowns in understanding how the climate system works and how it ultimately responds to increased greenhouse gas concentrations. While most of the basic principles of cloud mechanics are fairly well understood, what is not thoroughly understood are the quantitative ways clouds affect climate on a global scale.

Clouds both warm and cool the Earth and its lower atmosphere, and they are involved in a number of climate feedbacks—both positive and negative. (See Chapter 5.) Clouds contribute to warming by absorbing infrared radiation (heat) and releasing it toward the surface, and they cool by reflecting more incoming sunlight back out to space.

Furthermore, there are many different kinds of clouds, and their formation depends on a host of atmospheric conditions such as altitude, pressure, temperature, humidity, and vertical and horizontal air movement. They can be made of ice or water, with droplets of varying size, chemical composition, and density. These different kinds of clouds may have markedly different radiative effects.

Further, the different kinds of clouds evolve on many different scales of space and time. The scattered showers of summer may drench one neighborhood and leave dry the next one over, while a hurricane may produce a system of clouds that is hundreds of miles wide. Some clouds form and dissipate in minutes—their growth and dissolution visible to the naked eye—whereas others seem to last for hours.

These issues of complexity and scale make clouds especially hard to model, given the current state of the modeling technology and the limits of today's supercomputers. Cloud phenomena are too small and too fast to be simulated in the models used to represent possible global climate change effects.

Two extra complexities presented by clouds have to do with nucleation and chemistry. The microscopic droplets of liquid water that make up clouds tend to form and grow best around aerosols already in the air—dust, smoke, pollution, or preexisting liquid droplets. As a result, a variety of feedbacks and linkages involve both aerosols and clouds. For example, the shading/cooling effect of sulfate aerosols might actually be amplified if they stimulated cloud formation. But the washing out of those same sulfates from the atmosphere (as acid rain) might in turn be accelerated by the precipitation from those clouds—shortening the time over which those aerosols exert their cooling effect.

The **cloud albedo effect** describes the forcing effect from the change in cloud reflectivity—if the clouds become more reflective, more incoming sunlight is reflected to space, causing a cooling effect. The AR4, assessing results from a wide range of models, gives a wide range of values for the cloud albedo effect, from -0.22 W/m^2 to -1.85 W/m^2, but all models find a negative effect, indicating cooling. The models exhibit considerable differences in their handling of aerosols, cloud processes, and interactions between aerosols and cloud particles.

The uncertainty about the climate effects of clouds is magnified by the problem of feedbacks (see box in Chapter 3). Changing temperatures and humidities expected with climate change will result in changes to cloud cover—and those cloud cover changes will have further radiative effects. The problem is that it is not yet known whether the net effect of these feedbacks will be positive or negative—that is, whether they will have a cooling or warming effect on global average temperatures. While much progress has been made on including cloud feedbacks in climate models, the AR4 concludes that it's not yet possible to tell which climate model predicts the cloud feedback the best.

Further unknowns involve how cloud feedbacks and the net radiative effects of clouds will be distributed around the globe and through the year—regional and seasonal differences are quite possible. And finally, the uncertainty about cloud feedbacks is multiplied by separate uncertainties about the indirect cloud-forming effects of anthropogenic aerosols.

AEROSOLS

Figure 41. *An intense dust storm in March 2002 sent a massive plume of aerosols—small dust particles—outward over the Atlantic Ocean. Although much has been learned over the last decade about the role of aerosols in climate, many questions remain. Source: Jaques Descloitres, MODIS Rapid Response Team, NASA GSFC.*

Aerosols are microscopic solid particles or liquid droplets so light that they can be suspended in the air, temporarily, where they can alter the Earth's radiation budget and change regional climate. (For further background discussion of aerosols, see Chapter 4, The Human Effect).

Over the last decade, great progress has been made in our scientific understanding of the climatic effects of aerosols. The effect of aerosols on climate is more difficult to assess and predict than other forcers for a number of reasons. An initial problem is the uncertainty regarding estimates of the total amount of aerosols emitted to the atmosphere and remaining suspended in it at various altitudes. Best estimates of black carbon (one important aerosol), for example, are uncertain by a factor of two. Unlike the major greenhouse gases, which are well mixed globally and have long atmospheric lifetimes, aerosols remain in the atmosphere for a shorter time (typically days to weeks or months) and are therefore concentrated regionally.

There are other factors that also make it difficult to estimate actual amounts of aerosols. In some cases, the amount and type of the aerosols are determined by complex chemical processes occurring in the atmosphere as gases are transformed into solid or liquid particles. The transformation of gaseous sulfur dioxide to sulfate particles and sulfuric acid droplets is one example of this process.

In other cases (e.g., dust and gas from volcanoes or sea-salt particles from ocean-wave action) the aerosols are hard to measure and estimate because they are produced by highly variable and transient natural phenomena (e.g., many large eruptions one year, few the next). Finally, there are significant remaining uncertainties regarding the amounts of even comparatively well-understood aerosols coming from human activities—including sulfur particulates from human industrial combustion of fossil fuels and biomass burning from agriculture.

Even if scientists did know precisely the amounts of various aerosols in the atmosphere, there remain numerous uncertainties about just how they affect the Earth's radiative balance and climate.

Aerosol particles can reflect and scatter shorter, visible wavelengths of light or absorb longer, infrared wavelengths of radiated heat. The many different kinds of airborne particles have different radiative properties—scattering, absorbing, or both. These are known as **direct** aerosol effects. The AR4 reported that in total, all aerosols produce a direct forcing effect of -0.5 W/m^2, with a range of -0.9 to -0.1 W/m^2—a slight cooling effect (think of the fine particles as shading the Earth slightly from incoming sunlight). This range, even with a medium-low level of scientific understanding, represents the first ever estimate of the direct radiative effect of aerosols.

Studies continue to advance knowledge of types of aerosols in addition to sulfates—sea salt, humanly made and natural dust, smoke and soot from the burning of fossil fuels, smoke from natural and human burning of biomass, etc., but uncertainties remain. Further distinctions are being made between the **black carbon** component of smoke (sooty particles that are mostly carbon) and the **organic carbon** component, which consists of more complex and volatile carbon compounds. Much work remains to be done in measuring the atmospheric burden of each type of particle, learning the radiative properties of each, and assimilating this knowledge into models.

More difficult to assess are the **indirect** effects of aerosols. Indirect effects refer to changes in other physical properties, which in turn have measurable climate effects. Of particular interest is how aerosols change the properties of clouds. For example, aerosols may offer nuclei that encourage the condensation of cloud droplets (as in **cloud seeding**). Not only do they encourage cloud formation, but they may also change many other things—the cloud's size and longevity, the size and density of the droplets within it, and, most importantly, the radiative properties of the cloud.

Another kind of indirect effect is the opposite of the cloud-seeding effect—aerosols can also discourage the formation of precipitation from existing clouds. By increasing the number of droplets, they reduce average droplet size, inhibiting formation of droplets large enough to fall out of the cloud as rain. In discouraging the formation of precipitation from clouds, the aerosols can not only change precipitation patterns but also discourage the transformation of latent heat to sensible heat—cooling the atmosphere. All these effects are dependent on a number of variables, including the particular mixture of aerosols in a given cloud.

Because of their transient and regional nature, aerosol effects cannot be counted on to offset the anticipated longer-term global-scale warming trend.

The indirect effect of aerosols—primarily how they change the reflectivity of clouds—is even less well understood than the direct effect, but is currently estimated at -0.7 W/m^2, with a range of -1.8 to -0.3 W/m^2.

A final reminder: uncertainties like these are important to resolve—but they offer little support or comfort for those wishing to dismiss anthropogenic greenhouse warming. Because of their transient and regional nature, aerosol effects cannot be counted on to offset the anticipated longer-term, global-scale warming trend. It is important to note that some aerosols are predicted to decline in the coming decades, particularly as developing nations reduce air pollution, and although this is good for public health, it could effectively "unmask" greenhouse warming.

Carbon Budget

Carbon dioxide (CO_2) is the most important anthropogenic greenhouse gas, but there are still large unknowns about its global sources and sinks. The global carbon budget has been more precisely quantified in the last decade, and major advances have been made in distinguishing between CO_2 from fossil-fuel combustion and CO_2 from other sources, as well as in distinguishing between land and sea sinks. Nonetheless, remaining uncertainties limit the precision with which scientists can forecast the future growth of CO_2 concentrations.

One component of improvements in climate modeling in recent years is the inclusion of the carbon cycle. The AR4 reports that these **fully coupled** carbon cycle-climate models all predict a positive feedback (warming effect) from changes to the carbon cycle in a warmer world (although the strength of the feedback differs greatly among the models). This means that as the world warms, the land and oceans can be expected to take up less carbon, so that atmospheric concentrations of CO_2 would be that much higher.

The cycling of carbon (however balanced) between oceans and atmosphere, and between land and atmosphere dwarfs the human CO_2 disturbance, so more precise quantification of these parts of the carbon cycle is needed in order to develop more accurate projections about the growth of atmospheric CO_2. A lot more basic science and observation need to be done in order to fully understand the processes that cycle carbon through land and oceans.

One of the frontiers in knowledge of ocean carbon is largely a matter of physics and inorganic chemistry. The solubility (and thus rate of uptake) of atmospheric CO_2 in ocean water depends on physical properties, such as water temperature and the amount of CO_2 already dissolved in the water. The rate at which this "solubility pump" takes CO_2 out of the atmosphere is affected by a number of other factors as well, such as climate-related currents and the rate at which CO_2 or CO_2-rich water is transferred to lower depths. The oceans have absorbed about 30% of the CO_2 released by human activity since the start of the Industrial Revolution, and this uptake is measurably changing the chemistry of the upper layers of the ocean—with potentially dramatic consequences to ocean ecosystems and the marine food chain.

Other major unknowns in the carbon cycle are biological processes that remove or add CO_2 into the atmosphere. For example, CO_2 dissolved in ocean water is taken up by tiny algae and other marine plants, such as phytoplankton. These plants are the food that supports whole ecosystems of zooplankton (microscopic sea animals such as shrimp, for example). As the zooplankton die and decay, the calcium carbonate in their shells drifts down to the sea floor, moving carbon from surface waters down to deep and long-term storage in sediments and ultimately rock. But the rate at which this "biological pump" removes CO_2 is determined by a host of complex factors—light, current, dissolved nutrients, temperature, and ecological dynamics.

Likewise, much is still to be learned about the terrestrial carbon cycle involving soils, plants, and all living things. Both photosynthesis and decay are involved. Clearly the terrestrial system is large and complex, involving numerous linkages and feedbacks with climate and also with the concentrations of atmospheric gases. And soils and vegetation are clearly playing an important role in the carbon cycle—while about 45% of the carbon emitted since 1750 has remained in the atmosphere and 30% has been absorbed by the oceans, the remainder (about 25%) has been taken up by the terrestrial biosphere.

While major advances have been made in the modeling of all these difficult aspects of the carbon cycle, much more work remains to be done in understanding and quantifying the basic processes involved.

Figure 42. *A wildfire on the move in Washington's Paseyten Wilderness, July 1990. Wildfires are important global sources of greenhouse gases and aerosols to the atmosphere. According to the IPCC, fires "could increase or decrease depending on warming and precipitation patterns, possibly resulting under some circumstances in rapid losses of carbon." Source: Nick Sundt.*

Soils

Figure 43. *A Swedish botanist takes chamber measurements of CH_4 and CO_2 flux at a research site in Zackenberg National Park, Greenland. Source: Henning Thing, 1999, courtesy of Danish Polar Center/Polar Photos.*

Soils are a large and active reservoir of carbon—soils both remove carbon dioxide from the atmosphere and emit it into the atmosphere, making soils an important part of the carbon cycle. In addition, soils play several other roles related to climate.

Knowledge has grown as more soil types have been studied in relation to climate. Understanding of wetlands, from salt marshes to peat bogs, has helped fine-tune the global budgets of methane and carbon dioxide. Better appreciation of permafrost soils (frozen ground) as a reservoir of greenhouse gases has increased the need to understand how greenhouse warming might affect them and what feedbacks their thawing might cause. The top of the permafrost layer in the Arctic has increased in temperature by up to 3°C since the 1980s, and significant reductions in maximum extent and minimum depth have been observed. What this means for the release of large amounts of methane to the atmosphere—and for the potential for a runaway greenhouse feedback—is not well understood (see next section).

Land-use change generally refers to changes in vegetation, soils, and water resulting from human activities. Land-use change can significantly affect local climate by shifting radiation, changing cloudiness, and altering surface reflectivity and temperature. The IPCC judges that these processes have a low level of scientific understanding but concludes, "the impacts of land use change on climate are expected to be locally significant in some regions, but are small at the global scale in comparison with greenhouse gas warming."

Soil moisture is also a key factor in climate. One definition of drought is based on soil moisture, rather than precipitation. The level of soil moisture results from a number of factors, including the accumulation of moisture from many shorter-term rain or snow events.

However, if precipitation produces soil moisture, the reverse is also true: soil moisture produces precipitation. The evaporation of moisture from soil (or its **evapotranspiration** through plants) is in many inland regions a key source of the humidity that feeds precipitation. This is an example of a climate feedback, a tendency of weather patterns to persist—one reason that wet areas tend to stay wet, and dry areas tend to stay dry.

There are a number of other complexities related to soils. Soil moisture, for example, can encourage both plant growth and decomposition—influencing both sides of the atmospheric carbon balance. Evapotranspiration is not just a transfer of water to the air, but also a transfer of latent heat, which is just one of many ways in which soils regulate the transfer of heat from the earth to the atmosphere.

Climatologists have been aware of these complexities for a long time and have long included representations of soil moisture in their models. But the precision of models and the complexities they represent can always be improved. This is certainly true of efforts to understand soil processes and the way soils interact with vegetation and ecosystems, precipitation, and climate.

Better understanding of soil moisture as it relates to climate is important because this is one aspect of climate that affects people most directly—soil moisture determines the type, the success, or even the possibility of agriculture. Improvements of knowledge in this area will mean better projections of the regional impacts of climate change than are possible today.

In addition, soil moisture, soil type, and land contours are among the factors influencing the runoff of water into lakes and rivers. The erosion of soil and its transport to the sea via surface runoff is a small but significant factor in the global carbon cycle. Runoff of freshwater from land ultimately affects salinity distribution in the ocean. All in all, there is still much to learn about the connections between soil and climate.

Methane and Atmospheric Chemistry

Uncertainties about the global carbon budget are just some of the unknowns that remain with regard to greenhouse gases. The dynamics of methane are even less well understood than the dynamics of carbon dioxide.

The atmospheric concentration of methane in 2005 was more than double its pre-industrial value, which is unprecedented in the last 650,000 years. As stated clearly in the AR4, "Multiple lines of evidence confirm that the post-industrial rise in these gases [CO_2, CH_4, and N_2O] does not stem from natural mechanisms." In short, the increases observed in the last two centuries are due to human activities.

The growth rate of methane (that is, how fast the atmospheric concentration is increasing) was particularly high in the late 1970s and early 1980s, then decreased in the early 1990s, and finally hovered near zero from 1999 to 2005. A slight uptick in emissions was observed in 2007–08. (Note: this doesn't mean that emissions stopped—only that they stopped growing.) The AR4 finds that this slowdown in the CH_4 growth rate since 1993 is likely [> 66% probability] due to the atmosphere reaching a balance with near-constant emissions. The AR4 also notes, "*Insufficient understanding of the causes of recent variations in the CH_4 growth rate suggests large uncertainties in future projections for this gas in particular.*" (emphasis added)

Scientists are particularly concerned about a large release of methane from thawing permafrost, as the Arctic sees greater temperature increases than equatorial regions: "Observations also suggest increases in CH_4 released from northern peatlands that are experiencing permafrost melt, although the large scale magnitude of this effect is not well quantified," according to the AR4.

Similar large uncertainties surround more reactive greenhouse gases such as tropospheric ozone. The greenhouse impacts of tropospheric ozone are variable and localized, driven by all the dynamic atmospheric chemistry that contributes to the formation of smog. The chemistry becomes complicated because biologically produced emissions of ozone precursors (such as volatile organic compounds) depend on changes in temperature, humidity, and clouds. Climate change can also affect tropospheric ozone through changes in transport.

Ocean Currents

As discussed earlier, oceans are particularly important to global climate because of their role in the carbon cycle, and so scientists are interested in how the oceans are changing and will change under future warming. Circulation of ocean waters is an important part of the carbon cycle. Because of improvements in models, the important and complex role that oceans play in storing and redistributing the incoming solar energy toward the poles can now be described in more detail.

Note that these ocean currents not only flow on the surface, but also transfer water from the surface to the deep ocean and vice versa. These currents are driven by temperature as well as salinity (hence the term **thermohaline circulation**).

Although observations indicate clear changes in ocean waters, there is no clear evidence for changes in ocean circulation, according to the AR4.

One important ocean current is the Meridional Overturning Circulation (MOC for short). This current is caused by dense water at the poles that sinks to the deep oceans and then spreads over the equator. As the surface waters move toward the poles, the warm, salty water cools and becomes more dense. Vast volumes of ocean water are involved. This mechanism is of particular importance because there is evidence for a link between the MOC and abrupt climate changes in the last 120,000 years, but the exact relationship is unknown.

In short, the AR4 concludes that the MOC has changed significantly from year to year and from decade to decade through the end of the 20th century, but no clear pattern for a trend in the strength of the MOC has been found to date.

Because the exchange of carbon dioxide between ocean and atmosphere is so critical to an accurate portrayal of the Earth's carbon budget, better understanding of the effects of ocean currents on carbon exchange will be a key area of future research. Sudden or large shifts in ocean circulation could dramatically speed or slow the growth in the atmospheric concentration of CO_2.

The relative difficulty of predicting ocean behavior, along with its important linkages and feedbacks with the climate system (heat, humidity, carbon, etc.), make oceans an important source of uncertainty about the planet's climate future.

Ocean Chemistry

The chemistry of the ocean has changed, and continues to change, in a measurable way. From about 1750 to 1994, the total inorganic carbon content of the oceans has increased by 118 ± 19 GtC, and continues to rise. This represents some 30% of the total amount of carbon that has been released by human activity, which ended up in the oceans rather than in the atmosphere. The AR4 concludes that there is a >50% probability that the uptake of carbon by the oceans has decreased since 1994. This decline lines up well with what is expected based on how much the oceans can theoretically absorb, but the estimates are uncertain, making robust conclusions difficult.

The fact that the oceans have taken up vast amounts of emitted carbon dioxide is having a profound effect on the physical marine environment. As the carbon dioxide dissolves in seawater, it forms carbonic acid and increases the acidity of the water. The AR4 notes that direct observations of ocean pH over the last 20 years indicate a decline of about 0.02 units per decade—or an average of about 0.1 unit since 1750. This corresponds to a 26% increase in acidity levels. This change is already affecting various marine organisms that use calcium concentrate to construct their shells. The acidity levels expected by the end of the century (assuming no reductions in CO_2 emissions) would make oceans worldwide hostile to the growth of coral reefs—and may even cause them to dissolve. The impacts on other calcifiers, and the predators that rely on them, could have major implications for the marine food chain.

Extreme Weather

Much more than global average temperatures, people are naturally interested in extreme weather and climate events—hurricanes, tornadoes, floods, droughts, heat waves, blizzards, and so forth. Extreme weather has human impacts. Journalists know that these human impacts are what turn weather into news.

Simple statistics explains why we expect to see an increase in extreme events in a warmer climate. Figure 44 illustrates the concept that the probability of observing a particular event (such as a daily high temperature) can be represented as the familiar bell curve. The ends (or tails) of the curve represent infrequent, or extreme, events. As the mean shifts, the likelihood of such events changes—and previously extreme events become much more common.

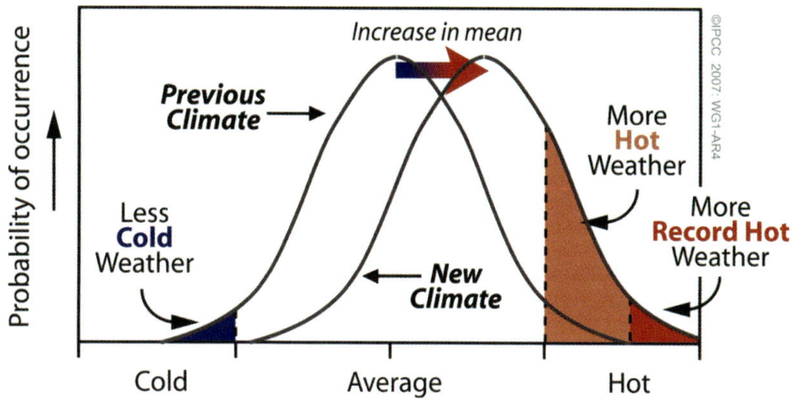

Figure 44. *This schematic shows how extreme weather becomes more common as the average increases. Extreme events (such as record temperatures) occur near the upper tails of the distribution. As shown, these extreme events are more probable as the mean temperature increases. Source: IPCC AR4 WGI, Box TS.5.*

The AR4 reports that observed changes in extreme temperature are consistent with the overall warming trend: "The warming of the climate is consistent with observed increases in the number of daily warm extremes, reductions in the number of daily cold extremes and reductions in the number of frost days at mid-latitudes."

One of the subtleties in explaining climate change is that, particularly for extreme events, we can only make projections about the *probability* of such events. It's difficult, if not impossible, to attribute a single weather event to climate change. While Hurricane Katrina in 2005 really represented a turning point in public perception of climate change, it's simply not correct to blame the storm entirely on climate change. But Katrina may be an example of an extreme event that could become more common in a warmer world.

Katrina may be an example of an extreme event that could become more common in a warmer world.

On a very simple conceptual level, such a hypothesis seems plausible. If warmer global climate were to bring warmer, late-summer sea-surface temperatures (SSTs) to the tropical Atlantic, hurricanes might well increase in frequency or intensity. But hurricanes, like many other extreme weather phenomena, are not nearly that simple.

A great deal of progress has been made on the question of hurricanes over the last decade. Observationally, the AR4 notes that intense tropical cyclone activity in the North Atlantic has increased since 1970, and that this increase correlates with increases in tropical SSTs. This observed increase, however, is greater than predicted by models for a future warmer climate. Model results indicate "the possibility of a decrease in the number of relatively weak hurricanes, and increased numbers of intense hurricanes." Much research remains to be done in order to answer this question definitively.

The IPCC reports that the observational basis for understanding extreme events has improved over the last decade, primarily because of longer data records. Daily temperature extremes and rainfall extremes, for example, have now been analyzed over most land areas. Improvements in models have provided greater insight into projection of extreme events, and climate detection and attribution studies have increasingly focused on extreme events. However, some events, such as hurricane intensity, suffer from inadequate data records and/or insufficient models, while others, such as future heat-wave occurrences, rely only on simple reasoning.

Phenomenon and direction of trend	Likelihood that trend occurred in late 20th century (typically post 1960)	Likelihood of a human contribution to observed trend	Likelihood of future trends based on projections for 21st century using SRES scenarios
Warmer and fewer cold days and nights over most land	*Very likely*	*Likely*	*Virtually certain*
Warmer and more frequent hot days and nights over most land areas	*Very likely*	*Likely* (nights)	*Virtually certain*
Warm spells / heat waves. Frequency increases over most land areas	*Likely*	*More likely than not*	*Very likely*
Heavy precipitation events. Frequency (or proportion of total rainfall from heavy falls) increases over most areas	*Likely*	*More likely than not*	*Very likely*
Area affected by droughts increases	*Likely* in many regions since 1970s	*More likely than not*	*Likely*
Intense tropical cyclone activity increases	*Likely* in some regions since 1970	*More likely than not*	*Likely*
Increased incidence of extreme high sea level (excludes tsunamis)	*Likely*	*More likely than not*	*Likely*

Table 4: *IPCC assessment of human influence on trends and projections for future extreme weather events. Source: IPCC AR4 WGI, Table SPM.2. Refer to the original table for details.*

Figure 45. *Source: Artville.*

Regional Impacts

Closely connected to the nature of extreme events are the projections of regional impacts of climate change. Scientific understanding of regional effects is becoming clearer as models improve. Primary improvements are better resolution, better simulation of regional effects, and a greater number of simulations. However, while **downscaling** (a particular method of translating global projections of climate change to the regional level) has matured over the last decade, there is much room for further development.

The AR4 makes the following conclusions regarding regional impacts, broadly defined:

- Warming is very likely [>90% probability] for all landmasses;
- Projected changes in precipitation are likely [>66% probability] or very likely [>90% probability for some regions;
- Confidence in projections of extreme events has improved over the TAR, particularly with regard to heat waves, heavy precipitation, and droughts.

In short, the IPCC says, "More specific information is now available across the regions of the world concerning the nature of future impacts, including for some places not covered in previous assessments," but confidence levels vary by region and by specific impact.

Human Behavior

Human influences are important to climate, and they are to some extent unpredictable. The human influences on climate are many—including combustion of fossil fuels, clearing of forests, agricultural practices that produce methane or cycle nitrogen, emissions of CFCs and other halocarbons, landfill methane, cement production, and burning of biomass in forests and fields. Predicting the future of these human activities is difficult, as is predicting the future impacts of these activities.

Fossil fuels provide one example. We have very good data about historical fossil fuel use, but projecting trends 50 or 100 years into the future is difficult. Oil consumption depends on many other things—accidents of geography, geopolitics, war, and economics. In fact, there are feedbacks between oil (or energy) prices and the cycle of economic growth. When fuels are perceived to be scarce or expensive, consumption of them decreases. To predict fossil fuel use 50 years from now, one would also need to be able to predict long-term economic growth. Economists are still not very good at this.

Technology, or more precisely technological change, is another example of the uncertainty regarding human influences on climate. The world today is full of technologies that seemed virtually impossible 100 or even 50 years ago: vaccines, antibiotics, cars and trucks, airplanes, synthetic fertilizers, high-yield crop varieties, pesticides, computers, the Internet, and so on. Future technological change has profound implications for future climate but

Figure 46. *Satellite image of Hurricane Katrina in the Gulf of Mexico, August 28, 2005. Credit NOAA.*

is hard to foresee or predict. No economic, population, or climate model can accurately account for such wild cards.

Population growth is also a basic determinant of the Earth's future climate and is to some degree unpredictable. Demographers and statisticians are certainly quite adept in their projections of future populations. But they would also be the first to admit the possibility of unknowns changing the planet's population future. Such unknowns include advances in contraceptive technology, miracle drug breakthroughs, newly emerging diseases like HIV/AIDS, other health catastrophes caused by disease or famine, government actions, social and cultural changes, religious movements, and so forth. Many of the ways that humans change climate are quite closely linked to population. The amounts of fossil-fuel combustion, biomass combustion, land clearing, and methane emissions from rice paddies and landfills, all change as population changes. The rate of global warming over the long term will depend partly on the rate of population growth, which is to a degree uncertain.

One big unknown is the degree to which—and the ways in which—humans will respond to climate change as it develops. **Adaptation** has gained increased attention, as policymakers begin to realize that society will have to prepare for unavoidable changes while avoiding the worst effects of climate change. Will people change cropping patterns and agricultural practices to capitalize on climate change? Will people migrate inland to escape rising seas? Will people respond with market choices to the changing abundance or scarcity of coal, oil, gas, biomass, and solar as fuels? Will people change energy use or land-use patterns?

The answers to some of these questions do not depend on government policies alone. To a considerable majority of the people on the planet, the driving concerns remain feeding, clothing, housing, and providing health care for their families. But climate change may still change people's perceptions of the world around them, and it remains to be seen how people will respond individually and collectively to changes in world and local climate.

CHAPTER 13: BRIEF GUIDE TO FALSE AND MISLEADING CONTRARIAN ARGUMENTS

First off, it's important for the scientific community—with the help of journalists—to reclaim the word "skeptics." In reality, all scientists are skeptical by nature. If a scientist makes a new claim or reaches a conclusion, her peers are initially skeptical, especially if the conclusion differs from current understanding—and it is incumbent upon her to support her conclusion with evidence. If her evidence holds up under the scrutiny of peers, and other experts are able to reproduce the results, the new conclusion gradually gains more widespread acceptance. Skepticism is the norm in scientific research. For that reason, and because of the strong scientific consensus that humans are affecting Earth's climate, it is more appropriate to refer to those who dispute the reality of climate change as "contrarians."

Part of the problem in communicating the science of climate change is, unfortunately, that the public is largely unaware of the scientific method. A researcher proposes an explanation for a phenomenon (called a hypothesis), devises an experiment to test that hypothesis, and then reaches a conclusion based upon those results. If the conclusion stands up to peer review, can be repeated by other researchers, and fits together with other evidence, eventually a theory is developed. If the theory continues to stand up over time, the scientific community generally accepts it.

However, it's important to understand that scientists very rarely (if ever) universally agree on a conclusion or theory. This is true in all branches of scientific research. Unfortunately, the public has the notion that any disagreement among experts means that no solid conclusion can be reached. The contrarians exploit this by casting doubt on the science of climate change. They continue to repeat long-since discredited arguments, simply to create doubt in the mind of the average citizen. Considering recent polls showing low public acceptance of the idea that humans are playing a large role in changing the climate, it would appear they have been largely successful. It is critical for journalists to understand that the "he-said, she-said" style of reporting (which may work well for political points of view) is simply inappropriate for reporting on the science of climate change: it means giving equal voice to a tiny minority of opinions.

It is critical for journalists to understand that the "he-said, she-said" style of reporting (which may work well for political points of view) is simply inappropriate for reporting on the science of climate change: it means giving equal voice to a tiny minority of opinions.

THE BOTTOM LINE ON CONTRARIAN MISINFORMATION

- Recent IPPC controversies have *not* altered the fundamental understanding of climate science or called into question that humans are the dominant cause of climate change.
- The much-publicized mistakes in the AR4 have been limited to very minor errors, and were not found in the WG1 (Physical Science Basis) report, upon which all the information in this guide is based.

Highly Publicized Misinformation

The last year has seen quite a bit of misinformation propagated by opponents of action on climate change, and it represents a good example of why accurate reporting of the science of climate change by journalists is absolutely critical. Both the IPCC and East Anglia e-mail "controversies" were blown out of all proportion, first by fringe elements and then by the mainstream press.

IPCC

Briefly, two mistakes from the AR4 received considerable press attention, and both appeared in the second volume, "Impacts, Adaptation, and Vulnerability" (Working Group II). But the fundamental scientific understanding of climate change is described in the first volume, "The Physical Science Basis," and no errors have yet sur-

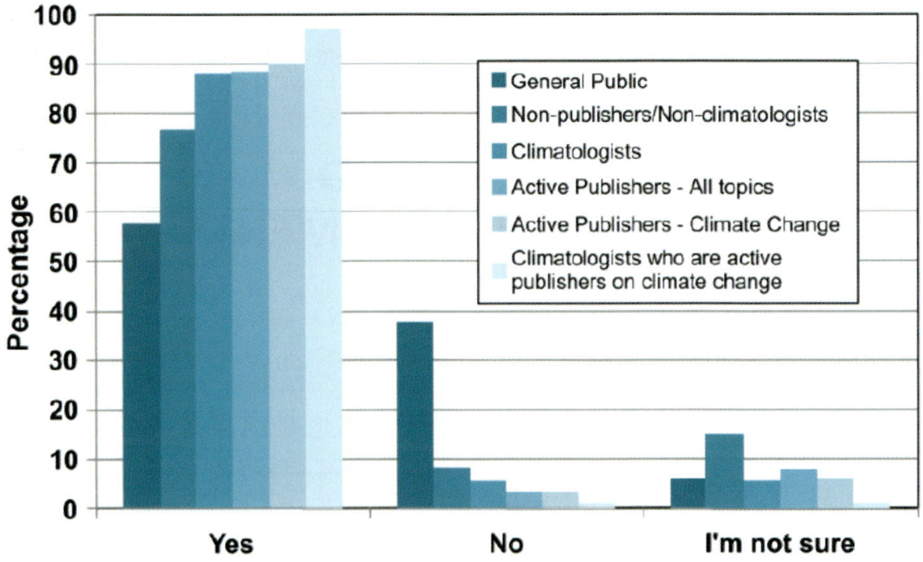

Figure 47. *Public Opinion and Profession. Source: Doran & Zimmerman (2009). EOS vol. 30, no. 3.*

faced in that volume—on which the information in this guide relies. Following is a very brief description of the IPCC controversies.

Himalayan Glaciers. The AR4's WGII volume contains a statement that the Himalayan Glaciers could disappear by 2035. This statement was apparently based on a non-peer-reviewed paper, and its inclusion is counter to IPCC guidelines. However, this single incorrect sentence does not take away from the robust and consistent statement appearing in the WGI volume:

> Climate change is expected to exacerbate current stresses on water resources from population growth and economic and land-use change, including urbanisation. On a regional scale, mountain snow pack, glaciers and small ice caps play a crucial role in freshwater availability. Widespread mass losses from glaciers and reductions in snow cover over recent decades are projected to accelerate throughout the 21st century, reducing water availability, hydropower potential, and changing seasonality of flows in regions supplied by meltwater from major mountain ranges (e.g. Hindu-Kush, Himalaya, Andes), where more than one-sixth of the world population currently lives.

Sea-Level Rise in the Netherlands. The IPCC was also criticized for allegedly overstating the threat of sea-level rise to the Netherlands. The WGII report included the following incorrect sentence: "The Netherlands is an example of a country highly susceptible to both sea level rise and river flooding because 55 percent of its territory is below sea level." It turns out that the figure includes land that is above sea level but susceptible to river flooding. The figure was originally obtained directly from the Dutch government, and the agency responsible, the Netherlands Environmental Assessment Agency, recently clarified that 26% of the land area is below sea level, and another 29% is susceptible to river flooding. At worst, the IPCC has received criticism over a statement it obtained directly from the Dutch government. In any event, the sentence was merely an example of the threat of sea-level rise, and in no way undermines the IPCC's conclusions on future sea-level rise.

In light of these errors, the United Nations and IPCC commissioned an independent review (http://reviewipcc.interacademycouncil.net/) of its policies and procedures by the InterAcademy Council, a consortium of science academies, including the National Academy of Sciences. The review recommended changes in administration and practices for the IPCC, but found IPCC's assessment processes to be sound.

HACKED E-MAILS

In November 2007 unknown persons hacked into the e-mail servers of the Climate Research Unit (CRU) at the University of East Anglia in England. The CRU is one of four organizations worldwide that independently gathers worldwide thermometer data to construct records of global temperature. The hackers sorted through and selected more than 1,000 e-mails that they posted on an anonymous FTP site in Russia. This action was illegal and is being investigated.

The vast majority of the e-mails were routine and unsuspicious. A dozen or two were impolite and give the *appearance* of controversy. To date, several independent investigations have concluded that the scientists involved in the e-mail exchanges acted ethically and properly, with the exception of possibly failing to respond to FOIA requests in a timely manner. Unfortunately, the "controversies" reported by the press were largely phrases that had been taken out of context. For example, one researcher wrote about using a "trick" to "hide the decline." A "trick" in scientist jargon is a clever or novel way of solving a problem—not an ulterior motive to suppress data, as was commonly reported in the press. A more in-depth description of the hacked e-mails was prepared by the Pew Center on Global Climate Change (http://www.pewclimate.org/docUploads/east-anglia-cru-hacked-emails-12-07-09.pdf). Three separate investigations have cleared the scientists of any wrongdoing.

RECORD SNOWFALLS

Two conditions are required for large amounts of snow: cold air and moist air. Cold air cannot hold large amounts of moisture, so heavy snowfalls usually occur when a cold, dry air mass collides with a warm, moist airmass. This condition is what occurred in February 2010, resulting in several feet of snow being dumped on the Washington, DC, metropolitan area. The warm air came from the Atlantic Ocean, where temperatures were higher, and the air contained larger amounts of moisture. These types of events are consistent with the science of climate change, as pointed out in the AR4:

> "Because precipitation comes mainly from weather systems that feed on the water vapour stored in the atmosphere, this has generally increased precipitation intensity and the risk of heavy rain and snow events."

Ironically, global warming can cause more frequent record snowfalls, as well as other extreme events. (See CCSP, 2008: *Weather and Climate Extremes in a Changing Climate. Regions of Focus: North America, Hawaii, Caribbean, and U.S., available at* http://www.globalchange.gov/publications/reports/scientific-assessments/saps/300). Although it seems counterintuitive, the record snow that struck many parts of the United States is actually consistent with current scientific understanding of climate change. This is an example of how the term "global warming" is really a misnomer—which is why most scientists refer to it as climate change, and why *New York Times* columnist Thomas Friedman refers to the term "global weirding." (See http://www.nytimes.com/2010/02/17/opinion/17friedman.html).

More importantly, it is critical, especially for journalists, to make a clear distinction between weather and climate. Weather forecasters are concerned with short-term phenomena (occurring on timescales of hours to days), while climatologists focus on longer-term changes (decades and longer). The general public (and unfortunately many television weathercasters) have the notion that since weather forecasts are typically not reliable ten days into the future, predictions of climate conditions in decades or a century from now are not believable. The analogy does not hold. One way to capture the distinction is that weather tells you if you need to bring an umbrella today, while climate determines whether you even own an umbrella.

The Misconception:	Recent global warming is caused by the sun.
The Reality:	The output of energy from the sun has been monitored by satellites for 30 years and has not increased during this period of rapid global warming.
The Misconception:	Climate has changed many times in the distant past, before humans began burning coal and oil, so the current warming cannot be caused by humans burning fossil fuels.
The Reality:	There are several drivers that cause climate to change, and some of the key drivers have both natural and human sources. Recent increases in global temperatures result mostly from higher levels of heat-trapping gases in the atmosphere, which have been increasing because of human activities.
The Misconception:	The last few years have been cooler, so global warming can't be real; or global warming stopped in 1998; or the world has been cooling for the past decade.
The Reality:	The climate is defined by long-term averages in global temperatures and other climate metrics, and those are still increasing.
The Misconception:	There is no scientific consensus on the existence or causes of global climate change.
The Reality:	A recent poll of earth scientists demonstrated that there is strong agreement that emissions of heat-trapping gases from the burning of fossil fuels make a significant contribution to global warming.
The Misconception:	Scientists predicted global cooling in the 1970s.
The Reality:	When the next ice age might occur became a topic of debate during the 1970s, but there was no concensus on the topic, and most of the debate was already focused on global warming.
The Misconception:	Atmospheric water vapor is the heat-trapping gas that is primarily responsible for global warming.
The Reality:	Water vapor is increasing in the atmosphere in response to rising CO_2 concentrations, amplifying the warming effect of human-made CO_2 emissions.

Table 5. *Misconceptions vs. Realities. Source: Pew Center on Climate Change, http://www.pewclimate.org/science-impacts/realities-vs-misconceptions*

Specific Contrarian Claims and Brief Rebuttals

Several resources exist that reliably address contrarian claims about climate change. In particular, the Pew Center on Global Climate Change has a useful website and report that delves in detail into frequent claims and soundbites from contrarians:

The website SkepticalScience.com is an excellent resource for addressing many contrarian claims.

In short, there is a strong motivator for scientists to prove the theory of anthropogenic global warming wrong—anyone who is able to do this would become as famous as Einstein! However, the burden of proof lies with the contrarian, who must demonstrate 1) another mechanism that could explain the multiple lines of evidence indicating that the Earth's climate is changing; and 2) why GHGs are not responsible, since basic physics indicates they should have a warming effect. So far, no one has been able to do this.

APPENDIX A: IPCC REPORTS PROCESS

The IPCC procedures specify the purpose of specific types of reports and procedures for preparation, review, acceptance, adoption, approval and publication of IPCC documents.

The IPCC's *Fourth Assessment Report*, published in 2007, includes four volumes:

- Climate Change 2007: The Physical Science Basis, by Working Group I
- Climate Change 2007: Impacts, Adaptation and Vulnerability, by Working Group II
- Climate Change 2007: Mitigation of Climate Change, by Working Group III
- Climate Change 2007: Synthesis Report

Each of the three Working Group reports includes a *Summary for Policymakers* and a *Technical Summary* in addition to the full report. The *Synthesis Report* includes a *Summary for Policymakers* and the full *Synthesis Report*. The *Synthesis Report* provides a high-level summary of the key findings from the three working groups. It uses cross-references extensively to demonstrate that conclusions are consistent with those reached in the three Working Group reports and with other documents approved and accepted by the IPCC.

Synthesis Reports, according to these procedures, "synthesize and integrate materials contained within the *Assessment Reports* and *Special Reports* and are written in a non-technical style suitable for policymakers and address a broad range of policy relevant but policy neutral questions." All of the reports are extensively reviewed by experts and government representatives.

The *Summaries for Policymakers* for each of the three Working Group reports undergo full line-by-line approval by a session of the sponsoring Working Group, to signify that they are consistent with the factual material contained in the full Working Group report.

Adoption of the three full Working Group Reports is a process of discussion and agreement section-by-section (not line-by-line) by the group. The reports are then reviewed and accepted by the IPCC. The *Summary for Policymakers* of the *Synthesis Report* is discussed and agreed upon line-by-line by the Panel and the full *Synthesis Report* is reviewed section-by-section.

The IPCC's complete procedures are available in *Procedures of the Preparation, Review, Approval, Acceptance, Adoption, and Publication of IPCC Reports*, which is available at http://www.ipcc.ch.

Likelihood Terminology	Likelihood of Occurrence or Outcome
Virtually certain	>99% probability
Extremely likely	>95% probability
Very likely	>90% probability
Likely	>66% probability
More likely than not	>50% probability
About as likely as not	33 to 66% probability
Unlikely	<33% probability
Very unlikely	<10% probability
Extremely unlikely	<5% probability
Exceptionally unlikely	<1% probability

Table 6. *The IPCC uses specific terminology to define the likelihood of projections and the confidence in scientific results. So, when the IPCC describes a result as "likely" or "unlikely," it has a very specific scientific meaning, as shown above. Source: IPCC, AR4, Box TS.*

APPENDIX B: GLOSSARY

Aerosols: small droplets or particles, larger than a molecule in size, that remain suspended in the atmosphere. They have a variety of sources—some natural, like dust storms and volcanic activity, and some caused by human activities, like fossil fuel and biomass burning. Note that there is no connection between these particulate aerosols and pressurized products also called aerosols.

Albedo: solar radiation reflectivity, with an albedo of one being the highest (reflecting all incoming light) and zero being the lowest (absorbing all light). Albedo varies depending on such factors as cloudiness, snow, and land cover. Because of its whiteness, snow typically has an albedo close to one; dull, black substances like charcoal have an albedo close to zero.

Anthropogenic: resulting from human activities.

Atmosphere: the mixture of gases that surrounds the Earth. It consists of about 79.1% nitrogen by volume, 20.9% oxygen, 0.036% carbon dioxide, and trace amounts of other gases. In addition, the atmosphere contains water vapor in the form of clouds and aerosols. Scientists divide the atmosphere into separate layers, according to mixing, chemical characteristics, and thermal properties. Those layers are the biosphere, troposphere, and stratosphere.

Atmosphere-Ocean General Circulation Models (AOGCMs): global climate models with coupled atmosphere and ocean components. They are highly complex and require a lot of computing power to run.

Biosphere: portion of the Earth that supports living organisms—on land (the terrestrial biosphere), in the oceans (marine biosphere), and in the atmosphere. It includes all ecosystems and living organisms, and also dead organic matter. Marine and terrestrial biospheres contribute to the atmosphere's composition.

Carbon Dioxide (CO_2): the principal greenhouse gas that affects the Earth's radiative balance. It occurs naturally and from such human activities as fossil-fuel burning, forest clearing, and other land-use changes.

Carbon sinks: reservoirs that take in and store more carbon than they release, thereby partially offsetting greenhouse gas emissions. Forests and oceans are two examples.

Carbon cycle: the flow of carbon (in its various forms, such as carbon dioxide) through the atmosphere, ocean, terrestrial biosphere, and lithosphere.

Chloroflorocarbons (CFCs): compounds containing chlorine, fluorine, and carbon. They are very stable in the troposphere but in the stratosphere are broken down by strong ultraviolet light, releasing chlorine atoms that then deplete stratospheric ozone. They act also as greenhouse gases, absorbing outgoing infrared radiation in the atmosphere. Common uses are as propellants, refrigerants, blowing agents (for producing foam), and solvents.

Climate: average weather pattern for a particular region and time period. It varies from place to place, depending on such factors as latitude, distance from the sea, vegetation, and the presence/absence of mountains. And it varies over time: by season, year, decade, or much longer periods.

Climate change: as defined by the IPCC, is any change in climate over time, whether resulting from natural causes or from human activity. These changes typically persist for decades or longer and may affect either the mean state of the climate or its variability. This definition differs from that of the UN Framework Convention on Climate Change (UNFCCC), which draws a distinction between climate change attributable to human activities, and climate variability attributable to natural causes.

Climate models: mathematical representations of the Earth's climate system and components and their processes and interactions. They are used as a research tool to study and simulate natural climate variability, and project the climate response to human activities (i.e., human-induced forcing). They are also used operationally for monthly, seasonal, and multi-year climate predictions. Climate models of varying complexity depict climate system components singly and in combination (coupled models).

Climate system: for purposes of the *Third Assessment Report*, the IPCC defines it as an interactive system consisting of five major components: the atmosphere, the hydrosphere, the cryosphere, the land surface, and the biosphere. The climate system continues to evolve over time, influenced by its own internal dynamics and by external forcings such as volcanic eruptions, solar variations, and human-induced forcings such as fossil fuel burning and land-use change.

Climate variability: climate changes that occur on time and spatial scales beyond those of individual weather events. Some of this variability is "forced" from outside the climate system itself—by things like anthropogenic greenhouse gases or solar variability. Other variability, such as oscillations in atmospheric-oceanic circulation, is internal to the climate system.

Cryosphere: the frozen part of the Earth's surface (on or beneath the surface of Earth and oceans), including the ice sheets of Greenland and Antarctica, continental glaciers and snow fields, sea ice, and permafrost. Its high reflectivity for solar radiation, low thermal conductivity, and large thermal inertia are important factors for the climate system.

Diurnal temperature range: difference between the maximum and minimum temperatures during a day.

Dobson Unit (DU): a measure of the total amount of ozone in a column of the atmosphere (total column ozone) from ground level to the top of the atmosphere, based on analysis of absorbed ultraviolet light. The number of Dobson units corresponds directly with the "thickness" of the ozone layer. While measurements vary widely according to time and place, a typical reading for a healthy polar ozone layer might be in the 300–450 Dobson unit range.

El Niño-Southern Oscillation (ENSO): pattern of climate/weather variation, known popularly as El Niño, that results from coupled atmosphere-ocean interactions, and recurs at two- to seven-year intervals. The ENSO pattern is driven partly by alternating warmer and cooler temperatures of the sea surface in the eastern and central tropical Pacific Ocean, which in turn are caused by changes in upwelling currents. It affects precipitation and temperatures over a large portion of the globe, with drastic consequences to human activities like farming and fishing, which depend on weather and ocean currents. The opposite of an El Niño event is called La Niña.

Eustatic sea-level change: average sea-level change caused by changes in water density or in the total mass of water.

Extreme weather events: weather phenomena that occur infrequently, such as droughts, heat waves, heavy rainfall, floods, hurricanes, and tornadoes.

General Circulation Model (GCM): a global, three-dimensional, computer climate-system model, which can be used to simulate human-induced change. GCMs are highly complex, depicting the interactions of such factors as greenhouse gas concentrations, clouds, annual and daily solar heating, mean ocean temperatures and ice boundaries, and the reflective and absorptive properties of atmospheric water vapor.

Glacial cycles: one of the most pronounced climate cycles, consisting of alternating ice ages and thaws—called glacial and interglacial periods. The last glacial cycle in human experience peaked about 20,000 years ago, with ice melting during the period 14,000–11,500 years ago. During the most recent million years or so (the epoch geologists call the Pleistocene), glacial cycles have come at fairly regular intervals of about 100,000 years.

Global surface temperature: average of near-surface air temperature over land, and sea-surface temperature. It is derived from sea-surface measurements, and land-surface readings taken 1.5 meters above the ground.

Global Warming Potential (GWP): the amount of global warming caused by a substance, expressed as the ratio of the warming caused by one substance relative to that caused by a similar mass of carbon dioxide over a given period of time.

Greenhouse effect: the effect produced as certain atmospheric gases allow incoming solar radiation to pass through to the Earth's surface but prevent the outgoing (infrared) radiation, which is re-radiated from Earth, from escaping into outer space. A certain amount of this occurs naturally, keeping the Earth's average temperature about 59 degrees Fahrenheit warmer than it otherwise would be.

Greenhouse gases: trace gases in the atmosphere that absorb and then emit infrared radiation in all directions, including downward to the Earth's surface, thus warming the lower atmosphere. The primary greenhouse gases in the Earth's atmosphere are carbon dioxide, nitrous oxide, methane, ozone, and water vapor.

Hydrosphere: climate system component comprising all liquid surface and subterranean water—includes both freshwater and salt water.

Halocarbons: compounds that combine carbon with either fluorine, chlorine, or bromine. These compounds can act as powerful greenhouse gases. Halocarbons containing chlorine and bromine also cause ozone depletion in the stratosphere. Included in the family of halocarbons are chlorofluorocarbons (CFCs) and the hydrochlorofluorocarbons (HCFCs).

Intergovernmental Panel on Climate Change (IPCC): organization established jointly by the United Nations Environmental Programme and the World Meteorological Organization in 1988 to assess information in the scientific and technical literature related to all significant components of the issue of climate change.

Kyoto Protocol to the United Nations Framework Convention on Climate Change (UNFCCC): international agreement, adopted in 1997 in Kyoto, Japan, wherein signatories agreed to reduce their anthropogenic greenhouse gas emissions (CO_2, CH_4, N_2O, HFCs, PFCs, and SF_6) to at least 5% below 1990 levels in the commitment period 2008 to 2012. Taking effect without ratification or approval by the United States.

Montreal Protocol on Substances that Deplete the Ozone Layer: international agreement, adopted in Montreal in 1987 and modified five times since then, that called for a freeze on production and use of halocarbons at 1986 levels by mid-1989, and over the next 10 years a reduction in CFC production by half. The U.S. and more than 180 other nations have ratified the agreement.

North Atlantic Oscillation (NAO): dominant pattern of northern, wintertime atmospheric circulation variability.

Ozone (O_3): bluish gas that is harmful to breathe, consisting of three bound atoms of oxygen. Nearly 90% of Earth's ozone is in the stratosphere, where it provides important benefits in absorbing harmful UV-B radiation, preventing most of it from reaching Earth's surface.

Ozone hole: observed depletion of the ozone layer over the Antarctic region that occurs yearly during the Southern Hemisphere spring. It is thought to be caused by the joint effects of chlorine and bromine compounds produced by people and meteorological conditions that are specific to the region.

Ozone layer: ozone in the stratosphere, where it occurs in its highest concentrations—roughly from 1 to 10 parts per million. This atmospheric zone lies between 15 and 50 kilometers above the Earth's surface, depending on latitude, season, and other factors. The term "ozone layer" is somewhat of a misnomer, since ozone does not occur in a flat layer in the atmosphere.

Pacific Decadal Oscillation: cyclic variations in sea-surface temperature in the northern Pacific—an example of natural climate variability. These occur on decadal timescales and affect the weather in places like the Pacific Northwest region of the United States.

Proxy record: historical record of climate-related variations obtained by examining tree rings, corals, ice cores, etc.

Radiative forcing: change in the balance between incoming solar radiation and outgoing infrared radiation. Causes include internal changes and external forcing, such as changes in solar output or carbon dioxide concentrations. Without any radiative forcing, solar radiation coming to the Earth would approximately equal the infrared radiation emitted from Earth. A positive forcing warms the Earth; a negative forcing cools it.

Solar radiation: energy from the sun—includes ultraviolet radiation, visible radiation, and infrared radiation.

Solar "11-year" Cycle: recurring pattern of solar output modulation, on scales of 9 to 13 years.

Stratosphere: zone of the atmosphere above the troposphere, extending from about 10 km (ranging from 9 km in high latitudes to 16 km in the tropics, on average) to about 50 km above Earth's surface. Commercial airplanes routinely fly in the lower stratosphere.

Thermohaline circulation (THC): global-scale overturning of the ocean driven by density differences arising from temperature and salinity effects. One of the best-known examples of thermohaline circulation is the Gulf Stream, a river of warmer, fresher, surface water that flows to the North Atlantic, where it gives up its heat and sinks, making much of Western Europe considerably warmer than it would otherwise be.

Trace gas: any one of the less common gases, together making up less than 1% of the Earth's atmosphere. Among these are carbon dioxide, water vapor, methane, nitrous oxide, ozone, and ammonia. Though small in absolute volume, they have significant effects on the Earth's weather and climate.

Troposphere: lowest part of the atmosphere, extending from the Earth's surface to about 10 km in altitude in mid-latitudes (ranging from 9 km in high latitudes to 16 km in the tropics, on average) where almost all "weather" phenomena occur. In the troposphere, temperatures generally decrease with altitude.

Ultraviolet (UV) radiation: that portion of the electromagnetic spectrum with wavelengths shorter than visible light. UV is commonly split into three bands: UV-A, UV-B, and UV-C. UV-A is not absorbed by ozone, UV-B is mostly absorbed by ozone, and UV-C is completely absorbed by ozone and normal oxygen.

United Nations Environment Programme (UNEP): the UN agency with a mission "to provide leadership and encourage partnership in caring for the environment by inspiring, informing, and enabling nations and peoples to improve their quality of life without compromising that of future generations."

United Nations Framework Convention on Climate Change (UNFCC): agreement signed at the 1992 Earth Summit in Rio de Janeiro by more than 150 countries and the European Community, with ultimate objective the "stabilization of greenhouse gas concentrations in the atmosphere at a level that would prevent dangerous anthropogenic interference with the climate system."

Weather: the fluctuating state of the atmosphere, characterized by temperature, wind, precipitation, clouds, and other weather elements.

World Meteorological Organization (WMO): Geneva-based, 185-member United Nations organization that provides "authoritative scientific voice on the state and behavior of the Earth's atmosphere and climate." Its stated purpose is to facilitate international cooperation in the establishment of networks of stations for making meteorological, hydrological, and other observations, and to promote the rapid exchange of meteorological information, the standardization of meteorological observations, and the uniform publication of observations and statistics.

APPENDIX C: UNITS OF MEASURE

UNITS OF MEASURE

DU — Dobson Unit, a measure of the total amount of ozone in a column of air (total column ozone) from ground level to the top of the atmosphere, based on spectral analysis of absorbed ultraviolet light. The number of Dobson units corresponds directly with the "thickness" of the ozone layer. While measurements vary widely according to time and place, a typical reading for a healthy, polar ozone layer might be in the 300–450 Dobson unit range.

GWP — Global Warming Potential, the amount of global warming caused by a substance, expressed as the ratio of the warming caused by one substance relative to that caused by a similar mass of carbon dioxide over a given period of time.

GtC — Gigatonnes of carbon (1 GtC = 1 PgC); one gigatonne equals 1 million metric tonnes; one tonne equals 1000 kilograms or approximately 2205 pounds.

PgC — Petagrams of carbon (1 PgC = 1 GtC); one petagram equals 1,000,000,000,000,000 (1015) grams.

PPBV — Parts per billion by volume, a measure of the concentration of a substance in the atmosphere; the greater the ppbv, the greater the percent of the atmosphere that consists of that substance.

PPMV — Parts per million by volume; see ppbv.

Tg — Teragram; equal to 1,000,000,000,000 (1012) grams.

W — Watt, a measure of energy output, equivalent to 1 joule per second; one megawatt equals one million watts.

W/m^2 — Watts per square meter, a unit used to measure the amount of solar energy/radiation that reaches the earth, or the earth's atmosphere.

ABBREVIATIONS OF CHEMICAL COMPOUNDS

CFC — chlorofluorocarbon
CO — carbon monoxide
CO_2 — carbon dioxide
CH_4 — methane
H_2O — water
N_2O — nitrous oxide
O — atomic oxygen (also called free oxygen or radical oxygen)
O_2 — oxygen
O_3 — ozone
OH — hydroxyl radical
SO_2 — sulfur dioxide

APPENDIX D: RESOURCES

Sources for Additional Information

Global Climate Change — Les changements climatiques

Official Canadian government climate change site. Includes press area.
Climate Change–Environment Canada
351 St. Joseph Blvd., Place Vincent Massey, 8th Floor
Gatineau, Quebec, K1A 0H3
Phone: (819) 997-2800, (800) 668-6767
Fax: (819) 994-1412
E-mail: media@ec.gc.ca
URL: http://climatechange.gc.ca/

Digital Library for Earth Science Education

A distributed community effort involving educators, students, and scientists working together to improve the quality, quantity, and efficiency of teaching and learning about the Earth system that is supported by the National Center for Atmospheric Research.
National Center for Atmospheric Research
PO Box 3000, Boulder, CO 80307-3000
303-497-1000 (no fax)
E-mail: support@dlese.org
URL: http://www.dlese.org

Intergovernmental Panel on Climate Change

The official site of the IPCC offers many official documents and reports, including its Third Assessment Report.
IPCC Secretariat, World Meteorological Organization Building
7 bis Avenue de la Paix, C.P. 2300, CH-211, Geneva 2, Switzerland
Phone: 4127308208
Fax: 41227308025
E-mail: ipcc_sec@gateway.wmo.ch
URL: http://www.ipcc.ch/

International Geosphere-Biosphere Programme

The International Geosphere-Biosphere Programme (IGBP) is an interdisciplinary scientific activity begun in 1986 and sponsored by the International Council for Science (ICSU). It seeks to provide an international, interdisciplinary framework for global-change science, to set priorities in the scientific problems approached, to establish consistency in the methods used to approach them, and to achieve compatibility of the resulting data.
IGBP Secretariat, The Royal Swedish Academy of Sciences
Box 50005, S-104 05, Stockholm, Sweden
Phone: 468166448
Fax: 468166405
E-mail: sec@igbp.kva.se
URL: http://www.igbp.kva.se/

United Nations Environment Programme

UNEP is the main UN environmental body and the organizational sponsor of many climate-related international efforts, including the IPCC and UNFCCC.
UNEP, PO Box 30552, Nairobi, Kenya
Phone: (254-20) 7621234
Fax: (254-20) 7624489/90
E-mail: unepinfo@unep.org
URL: http://www.unep.org

United Nations Framework Convention on Climate Change (UNFCCC)

The official site of the UNFCCC. The Framework Convention is the international treaty under which nations seek to limit greenhouse warming. The role of the FCCC secretariat is diplomatic rather than scientific. Site is searchable.
UN–FCCC, Haus Carstanjen, Martin-Luther-King-Strasse 8, D- 53175 Bonn, Germany
Phone: 492288151000
Fax: 492288151999
E-mail: secretariat@unfccc.int
URL: http://unfccc.int

The Met Office (U.K.)/Hadley Centre for Climate Prediction and Change
The Met Office is the national weather service for the United Kingdom. The Hadley Centre is part of the Office.
The Met Office, FitzRoy Road, Exeter
Devon, EX1 3PB, United Kingdom
Phone: 441392885680
Fax: 441392885681
Press Office: 01344856655
E-mail: enquiries@metoffice.gov.uk
URL: http://www.meto.gov.uk/index.html

World Meteorological Organization
This UN subsidiary coordinates international networks of meteorological observations and exchange of data and research.
7 bis Avenue de la Paix, CP 2300 – 1211, Geneva 2, Switzerland
Phone: 410227308111
Fax: 410227308181
E-mail: wmo@wmo.int
URL: http://www.wmo.int/pages/index_en.html

U.S. Federal

U.S. Environmental Protection Agency

EPA Global Warming Site
A good overall resource on climate change.
URL: http://www.epa.gov/climatechange/

EPA: Ozone Depletion Site
This Web site, maintained by EPA's Stratospheric Protection Division, contains information about the science of ozone depletion, U.S. laws and regulations to protect the ozone layer, international treaties, CFC substitutes, methyl bromide, the UV index, and other topics.
EPA Ozone Protection Hotline: (800) 296-1996
Fax: (301) 231-6377
URL: http://www.epa.gov/ozone/strathome.html

Department of Energy

DOE: Energy Information Administration — Greenhouse Gases
Information on estimated emissions of greenhouse gases in the United States and worldwide. The EIA compiles and distributes energy statistics.
URL: http://www.eia.gov/environment/

DOE: Oak Ridge National Laboratory — Climate Change Prediction Program
This is DOE's program to develop, compare, and use computer models to simulate climate change.
URL: http://www.epm.ornl.gov/chammp/chammp.html

DOE: Carbon Dioxide Information Analysis Center (CDIAC)
This center is devoted to research and information about the global carbon cycle and its climate implications. Run by the Energy Department's Oak Ridge National Laboratory.
Phone: (865) 574-0390
Fax: (865) 574-2232
URL: http://cdiac.ornl.gov/

DOE: Atmospheric Radiation Measurement (ARM) Program
This research program aims at measuring and understanding the flows of solar and thermal energy through Earth's climate system.
ARM toll-free information number: (888) 276-3282
URL: http://www.arm.gov/

National Aeronautics and Space Administration

NASA "Newsroom" page
URL: http://www.nasa.gov/news

NASA Television (NTV)
Offers real-time coverage of Agency activities and missions for news media and others. The schedule of upcoming video is downloadable from ftp://ftp.hq.nasa.gov/pub/pao/tv-advisory/nasa-tv.txt.
URL: http://www.nasa.gov/ntv/

NASA's photo, video, and audio galleries
Public Affairs Office, Room 5N22, NASA Headquarters, Washington, DC 20546
Phone: (202) 358-1730
URL: http://www.nasa.gov/gallery/index.html

NASA Ames Research Center, Earth Science Division
This NASA unit conducts research in a number of areas related to climate: atmospheric chemistry and physics, clouds, aerosols, stratospheric ozone, etc., using airborne instruments.
NASA Ames Research Center, Earth Science Division, Mail Stop 245-4, Moffett Field, CA 94035-1000
Phone: (650) 650-9000
URL: http://geo.arc.nasa.gov/

NASA Earth Observing System Home Page
NASA's Earth Observing System (EOS) uses a network of satellites carrying advanced instruments to make long-term global observations of the land surface, biosphere, solid Earth, atmosphere, and oceans. It also includes a system for collecting, storing, and analyzing the huge volume of data that will be generated.
URL: http://eospso.gsfc.nasa.gov

The Global Change Master Directory
NASA's Global Change Master Directory (GCMD) is a comprehensive directory of descriptions of data sets of relevance to global change research. The GCMD data-base includes descriptions of data sets covering climate change, the biosphere, hydrosphere and oceans, geology, geography, and human dimensions of global change.

GCMD User Support Office
Science Systems and Applications, Inc., 10210 Greenbelt Road, Suite 500, Lanham, MD 20706
Phone: (301) 867-2085
E-mail: gcmduso@gcmd.nasa.gov
URL: http://gcmd.gsfc.nasa.gov/

NASA: Goddard Space Flight Center — Climate and Radiation Branch
This NASA unit conducts a full range of climate research, including remote sensing and modeling — with special emphasis on aerosols and clouds.
Phone: (301) 614-6183
Fax: (301) 614-6307
URL: http://climate.gsfc.nasa.gov/

NASA Global Change Data Center
Data products from NASA and the Mission to Planet Earth.
URL: http://science.gsfc.nasa.gov/sed/index.cfm?fuseAction=home.main&&navOrgCode=610.2

National Oceanic and Atmospheric Administration

NOAA Public and Constituent Affairs
NOAA Public & Constituent Affairs, Constitution Ave. NW, Room 5128, Washington, DC 20230
Phone: (202) 482-6090
Fax: (202) 482-3154
URL: http://www.publicaffairs.noaa.gov/

National Climatic Data Center (NCDC)
NOAA's NCDC includes archive of weather data and publications analyzing weather variability, extreme weather, and climate trends. NCDC partners with regional climate centers, and state climatologists.

Climate Services Branch, National Climatic Data Center
151 Patton Ave., Rm. 468, Asheville, NC, 28801-5001
Phone: (828) 271-4800
Fax: (828) 271-4876
E-mail: ncdc.info@noaa.gov
URL: http://lwf.ncdc.noaa.gov/oa/ncdc.html

NOAA Office of Global Programs
NOAA Office of Global Programs, 1100 Wayne Avenue, Suite 1210, Silver Spring, MD 20910
Phone: (301) 427-2089
Fax: (301) 427-2073
Phone: (301) 427-2089 x137
URL: http://www.ogp.noaa.gov/

NOAA Central Library
Silver Spring Metro Center Building 3 (SSMC3), 2nd Floor, 1315 East-West Highway, Silver Spring, MD 20910
Reference Desk: (301) 713-2600, ext. 124
Fax: (301) 713-4599
E-mail: Library.Reference@noaa.gov
URL: http://www.lib.noaa.gov/

NOAA Central Library — Online Free Photo Collection
This collection of more than 10,000 digitized images related to the atmosphere and oceans is copyright-free and available online.
Phone: (301) 713-2600 x115
E-mail: photolibrary@noaa.gov
URL: http://www.photolib.noaa.gov/

Geophysical Fluid Dynamics Laboratory
NOAA's climate modeling research center.
Geophysical Fluid Dynamics Laboratory/NOAA
Princeton University Forrestal Campus
201 Forrestal Rd., Princeton, NJ 08540-6649
Phone: (609) 452-6500
Fax: (609) 987-5063
URL: http://www.gfdl.gov/

NOAA Pacific Marine Environmental Laboratory (PMEL: Seattle, WA)
This major center of NOAA oceanographic research maintains the Tropical Atmosphere Ocean (TAO) Array—a system of buoys spanning the tropical Pacific, which is one of the major gauges of El Niño activity.
NOAA R/PMEL,
7600 Sand Point Way NE, Seattle, WA 98115
Phone: (206) 526-6239
Fax: (206) 526-6815
URL: http://www.pmel.noaa.gov/

NOAA National Weather Service
The National Weather Service contains a host of resources relevant to climate change and variability.
National Weather Service
1325 East-West Highway, Silver Spring, MD 20910
Phone: (301) 713-0622
URL: http://www.nws.noaa.gov/pa/

NOAA-NCEP Climate Prediction Center Site
Offers short-term climate (or long-term weather).
NOAA/NWS, National Centers for Environmental Prediction, 5200 Auth Road, Camp Springs, MD 20746
Phone: (301) 763-8000, ext. 7163
E-mail: jana.goldman@noaa.gov
URL: http://www.cpc.ncep.noaa.gov/
Within NOAA's Office of Oceanic and Atmospheric Research (OAR), several different labs, programs, and institutes conduct research on global warming, ozone depletion, weather, and oceans.
Phone: (301) 713-2458
URL: http://www.oar.noaa.gov/

The Cooperative Institute for Research in Environmental Sciences (CIRES)
A cooperative effort of the University of Colorado and NOAA. It is involved in studying basic and applied problems associated with the physics and chemistry of the solid Earth and its atmosphere, oceans, and cryosphere.
CIRES, University of Colorado
Campus Box 216, Boulder, CO 80309-0216
Phone: (303) 492-1143
Fax: (303) 492-1149
E-mail: info@cires.colorado.edu
URL: http://cires.colorado.edu/

NOAA–Earth Sciences Research Laboratory, Chemical Sciences Division (CSD)
CSD's mission is to discover, understand, and quantify the chemical reactions and related processes that affect the Earth's atmosphere by focusing on climate change, air quality, and stratospheric ozone.
NOAA Earth System Research Laboratory
Chemical Sciences Division, 325 Broadway R/CSD, Boulder, CO 80305-3337
URL: http://www.esrl.noaa.gov/csd/

NOAA–Earth Sciences Research Laboratory, Physical Sciences Division (PSD)
PSD conducts weather and climate research to observe and understand Earth's physical environment, and to improve weather and climate predictions on global-to-local scales.
NOAA Earth System Research Laboratory
Physical Sciences Division, 325 Broadway R/PSD, Boulder, CO 80305-3328
Phone: (303) 497-4233
URL: http://www.esrl.noaa.gov/psd/

NOAA–Earth Sciences Research Laboratory, Global Monitoring Division (GMD)
GMD conducts observations and research related to source and sink strengths, trends and global distributions of atmospheric constituents that are capable of forcing climate change, that may cause depletion of the global ozone layer, and that affect baseline air quality.
NOAA Earth System Research Laboratory
Global Monitoring Division, 325 Broadway R/GMD1, Boulder, CO 80305-3337
URL: http://www.esrl.noaa.gov/gmd/

Other Federal and National Sources

U.S. Global Change Research Program
This interagency program coordinates federal research on global change in many agencies and universities. It was formally mandated by Congress in a 1990 law, and is now under the National Science and Technology Council's Committee on Environment and Natural Resources.
U.S. Global Change Research Program, Suite 250, 1717 Pennsylvania Ave. NW, Washington, DC 20006
Phone: (202) 223-6262
Fax: (202) 223-3065
URL: http://www.globalchange.gov/

U.S. Global Change Research Information Office
The site of the information arm of the U.S. Global Change Research Program provides access to many of the most important products of federally sponsored research.
U.S. Global Change Research Information Office
Suite 250, 1717 Pennsylvania Ave. NW
Washington, DC 20006
Phone: (202) 223-6262
Fax: (202) 223-3065
Email: information@gcrio.org
URL: http://www.gcrio.org/

U.S. State Department, Bureau of Oceans and International Environmental and Scientific Affairs
Office of Press Relations (Room 2109), U.S. Department of State, Washington, DC 20520-6180
Phone: (202) 647-2492 (press queries)
E-mail: askpublicaffairs@state.gov
Press briefings on Web: http://www.state.gov/r/pa/prs/dpb/
Press briefings via fax on demand: (202) 736-7720.

USDA National Water and Climate Center
This site, maintained by USDA's Natural Resources Conservation Service, includes information on agriculture's vulnerability to climate variability like drought and mitigation measures.
Natural Resources Conservation Service, National Water and Climate Center, 1201 NE Lloyd Blvd., Suite 802 Portland, Oregon 97232-1274
E-mail: info@wcc.nrcs.usda.gov
URL: http://www.wcc.nrcs.usda.gov/

Regional, State, and Local

Regional Climate Centers
This site contains links to the six U.S. regional climate centers.
URL: http://www.wrcc.dri.edu/rcc.html

American Association of State Climatologists
Includes a listing of all state climatologists with contact information.
URL: http://www.stateclimate.org/l

International Council for Local Environmental Initiatives' Climate Program
International Council for Local Environmental Initiatives' Climate Program consists of more than 300 local governments worldwide, including over 50 U.S. cities and counties, that seek to tackle climate change.
URL: http://www.iclei.org/ccp/

'Contrarians'

GlobalWarming.org
Globalwarming.org is the blog of the Cooler Heads Coalition, an ad hoc coalition of more than two dozen free market and conservative non-profit groups in the United States and abroad that question global warming alarmism and oppose energy-rationing policies.
URL: http://www.globalwarming.org/

The Science & Environmental Policy Project
The Science & Environmental Policy Project was founded by atmospheric physicist S. Fred Singer. The site has a major collection of the arguments of greenhouse skeptics.
URL: http://www.sepp.org/

World Climate Report
The electronic edition of the flagship periodical of the global warming contrarians, founded by University of Virginia's Pat Michaels, who is a fellow with the libertarian Cato Institute.
URL: http://www.worldclimatereport.com/

George C. Marshall Institute
This organization, originally focused on nuclear defense issues, says it wants to bring sound science to bear on public policy. Its interest has turned to global warming.
The George C. Marshall Institute, 1601 North Kent Street, Suite 802, Arlington, VA 22209
Phone: (571) 970-3180
Fax: (571) 970-3192
E-mail: info@marshall.org
URL: http://www.marshall.org/

Junk Science Page
An iconoclastic review of science stories with a skeptical viewpoint. It is published by Steven J. Milloy, an adjunct scholar at the Cato Institute.
URL: http://www.junkscience.com/

Center for the Study of Carbon Dioxide and Global Change
This group is run by Arizona State University geographer Craig D. Idso and botanist Keith E. Idso. It is less ideology-driven than other contrarian organizations, and its strength is in its focus on the role of plants in the global carbon cycle.
Center for the Study of Carbon Dioxide and Global Change, PO Box 25697, Tempe, AZ 85285-5697
Phone: (480) 996-3719
Fax: (480) 996-0758
E-mail: contactus@co2science.org
URL: http://www.co2science.org/

Environmental Organizations and Other NGOs

CIESIN — Center for International Earth Science Information Network
CIESIN is part of Columbia University's Earth Institute. CIESIN provides information to help scientists, decisionmakers, and the public better understand the changing relationship between human beings and the environment.
CIESIN, 61 Route 9W, PO Box 1000, Palisades, NY 10964
Phone: (845) 365-8988
Fax: (845) 365-8922
Washington, DC Office
Phone: (202) 419-3467
Fax: (202) 488-8679
E-mail: info@ciesin.columbia.edu
URL: http://www.ciesin.org/

Climate Action Network International
CAN International is a worldwide network of roughly 500 nongovernmental organizations working to promote government and individual action to limit human-induced climate change to ecologically sustainable levels.
Climate Action Network International
1810 16th St. NW, Washington, DC 20009
Phone: (202) 621-6309
Fax: (202) 536-5503
URL: http://www.climatenetwork.org/

Environmental Defense Fund
257 Park Avenue South, New York, NY 10010
Phone: (212) 505-2100
E-mail: press@environmentaldefense.org
URL: http://www.edf.org

Institute for Energy and Environmental Research
6935 Laurel Ave., Suite 201, Takoma Park, MD, 20912
Phone: (301) 270-5500
E-mail: info@ieer.org
URL: http://www.ieer.org

Natural Resources Defense Council
40 West 20th Street , New York, NY 10011
Phone: (212) 727-2700
E-mail: nrdcinfo@nrdc.org
URL: http://www.nrdc.org

Pew Center on Global Climate Change
2101 Wilson Blvd., Suite 550, Arlington, VA 22201
Phone: (703) 516-4146
Fax: (703) 841-1422
URL: http://www.pewclimate.org/

Sierra Club
85 Second Street, 2nd Floor , San Francisco, CA 94105
Phone: (415) 977-5500
E-mail: media.team@sierraclub.org
URL: www.sierraclub.org

Renewable Energy Policy Project (REPP)
REPP supports the development of renewable energy technology through policy research.
1612 K Street NW, Suite 202, Washington, DC 20006
Phone: (202) 293-2898
Fax: (202) 293-5857
E-mail: gsterzinger@repp.org
URL: http://www.repp.org/

SEDAC: Stratospheric Ozone and Human Health Project
The Socioeconomic Data and Applications Center is an element of NASA's Earth Observing System Data and Information System (EOSDIS). It is a good starting point for background on climate modeling and the health effects of increased UV exposure from ozone depletion, although the site is no longer being updated.
URL: http://sedac.ciesin.org/ozone/

World Resources Institute
10 G Street NE, Suite 800, Washington, DC 20002
Phone: (202) 729-7600
E-mail: moko@wri.org
URL: http://www.wri.org

Worldwatch Institute
1776 Massachusetts Ave. NW, Washington, DC 20036
Phone: (202) 452-1999
E-mail: worldwatch@worldwatch.org
URL: http://www.worldwatch.org